From *Logos* to *Bios*

Evolutionary Theory in Light
of Plato, Aristotle & Neoplatonism

From *Logos* to *Bios*

Evolutionary Theory
in Light of
Plato, Aristotle *&* Neoplatonism

✛

Wynand de Beer

First published
by Angelico Press 2018
© Wynand de Beer 2018

All rights reserved

No part of this book may be reproduced or transmitted,
in any form or by any means, without permission.

For information, address:
Angelico Press
169 Monitor St.
Brooklyn, NY 11222
angelicopress.com

978-1-62138-344-4 (pbk)
978-1-62138-345-1 (cloth)
978-1-62138-346-8 (ebook)

Cover design: Michael Schrauzer

CONTENTS

Acknowledgments i

Introduction 1

1 • *The Metaphysical Concept of Evolution* 5

2 • *Aspects of Hellenic Cosmology* 14
Presocratic Cosmology—The Cosmology of Plato—The Metaphysics of Aristotle—Neoplatonic Cosmology

3 • *Hellenic Philosophy of Life, or Bio-Philosophy* 39
Plato—Aristotle—Form—Reproduction and Growth—Soul—Aristotle's Scale of Nature—Aristotle's Teleology—The Neoplatonists—The Great Chain of Being

4 • *Form and Transformation* 71
Organic Form—Theory of Transformations—Variations Based on Rectangular Co-Ordinates—Three-Dimensional Co-Ordinate Systems—Relevance—The Limits of Transformation

5 • *The Modern Theory of Evolution* 98
Darwin and Natural Selection—The Rise of Genetics—Dobzhansky and the Modern Evolutionary Synthesis—Gould and Punctuated Equilibrium—Criticism—Variation *versus* Speciation—The Origin of Organic Complexity—Sudden Speciation Followed by Stasis—The Analogy of Artificial Breeding—The Improbability of Contingency—From Simple to Complex—The Struggle for Existence—Utility *versus* Teleology—Progress and Evolution—Materialism *versus* Levels of Being—Transformation *versus* Manifestation—Conclusion

6 · *Evolution According to Natural Law* 146
The Origin of Adaptations in Organisms—Law, Chance, and Design—The Role of Selection—The Course of Evolution—The Origin of Genes—The Orthogenetic Formation of Characters—The Role of the Geographical Landscape—The Polyphyletic Origins of Similar Forms—The Formation of New Species—The Limits Between Species—Mutation and Speciation—The Laws of Evolution—Darwinism *versus* Nomogenesis—Paleontological Confirmation of Mass Mutations—Biochemical Confirmation of Evolution Determined by Law

7 · *Directed Evolution* 198
Development of the Theory—Constraints on Variation—Biochemical Evidence—Evolution by Means of Adaptive Mutation—Organic Plenitude—The Constraints of Functional, Integrative, and Irreducible Complexity

8 · *Convergent Evolution* 235
Convergence of Internal Characters—Convergence of External Characters—Convergence of Psychical Phenomena—Convergence and Evolution

Conclusion 258

Bibliography 265

Index 273

Acknowledgments

Since the central argument of this book is based on my doctoral thesis, I should first of all like to thank the University of South Africa for providing me with a scholarship. A debt of gratitude is owed in particular to my promoter, Danie Goosen, for his wise guidance during the years of my research. I also thank the examiners of my thesis at the Universities of South Africa, Kentucky, and Portsmouth for their helpful criticisms and suggestions. It has indeed been an international project.

I furthermore wish to express gratitude to my friend Robert Proctor for introducing me to the work of René Guénon, Ananda Coomaraswamy, Frithjof Schuon and other Traditionalist authors, including the traditional understanding of evolution as further elaborated in this book. In addition, I should like to thank my friends Andrew Bassil and Colin Gemmell for stimulating conversations with regards to evolutionary biology.

Gratitude is also due to John Riess at Angelico Press for his advice throughout the publication process. Last, but not the least, I thank my editor Marie Hansen, whose philosophical acumen and literary proficiency enabled her to transform my academic manuscript into a readable and accessible book. By the nature of things, any errors of fact or judgement remain my own.

Dedicated to my parents,
ALBERTUS and JOHANNA DE BEER

Introduction

THE EVOLUTIONARY THEORY of the origin of life on Earth has been one of the most contentious intellectual issues in the natural sciences and beyond ever since it was systematically laid out in Charles Darwin's 1859 publication, *On the Origin of Species by Means of Natural Selection*. The consequent modern theory popularly known as Darwinism incorporates elements of both scientific theory and ideology, and forms the biological foundation of our current scientific framework. This framework has its roots in thinkers such as Galileo, Descartes, and Bacon, and rests upon the twin pillars of materialism and mechanism, which purport to explain reality solely in terms of quantitative laws. With the so-called Age of Enlightenment, an ever-increasing number of people in the Western world came to hold that the natural sciences provided a more reliable source of knowledge than religious revelation or the teachings of philosophers, and the idea that the universe could be explained purely as the result of autonomous natural laws came to be generally accepted.

By the nineteenth century, the traditional notion of a Great Chain of Being, according to which all levels of reality are related in a cosmic hierarchy with God at the summit, had become horizontalized and temporalized. This ontological deprivation provided the fertile ground needed for Darwin's theory of evolution to take root as the most obvious explanation for biodiversity, ousting belief in divine creation. Darwin's theory soon became dogma, replacing those of religion for those who had lost their faith in divine causality. It is precisely as dogma that the Darwinian theory of evolution has been able to survive to this day in the face of contrary evidence, especially from molecular biology and paleontology. In the words of the Sufi philosopher Seyyed Hossein Nasr, Darwinism has thus come to function as "a convenient philosophical and rationalistic scheme to enable man to create the illusion of a purely closed Universe around himself."[1]

Hellenic philosophy proves relevant to the debate about evolution because it provides an illuminating alternative to the unsatisfying dichotomy that from the outset has dominated the ongoing debate concerning the origins of biodiversity. On the one hand are those who reject any notion of organic evolution on supposedly religious grounds, and on the other hand those who support the most extreme notions of materialistic and mechanistic evolution, respectively associated in the popular mind with religious fundamentalists and atheist scientists. Opposed to both, the philosopher Alfred North Whitehead remarks that the apparent conflict between science and religion is attributable to the deficiencies of both scientific materialism and super-naturalistic theism,[2] and the Ceylonese intellectual Ananda Coomaraswamy writes that "a real conflict of science with religion is unimaginable; the actual conflicts are always of scientists ignorant of religious philosophy with fundamentalists who maintain that the truth of their myth is historical."[3] Hellenic philosophy, too, with its comprehensive integration of the metaphysical and the physical, stands in stark contrast to both of these positions, avoiding the conceptual Scylla and Charybdis of atheistic materialism and religious supernaturalism.

We contend that the modern evolutionary theory, Darwinism in all of its permutations, is deficient on both metaphysical and empirical grounds. Metaphysically, it is implausible due to its rejection of formal and final causality, and on the empirical level, it is unable to satisfactorily explain macro-evolution, including such phenomena as the sudden appearance of new phyla, classes, orders, and families, as well as the appearance of complex organic structures such as the

---

1. Nasr, Seyyed Hossein. "Progress and Evolution. A Reappraisal from the Traditional Perspective," in *Parabola*, Volume VI, Number 2 (1981): 49–50.

2. Griffin, David Ray, "Interpreting Science from the Standpoint of Whiteheadian Process Philosophy," in *The Oxford Handbook of Religion and Science*, eds. Philip Clayton and Zachary Simpson (Oxford: Oxford University Press, 2008), 454–55.

3. Coomaraswamy, Ananda, *What is Civilization? And Other Essays* (Ipswich: Golgonooza Press, 1989), 72.

Introduction

brain, eye, feather or spider-web. In contrast to micro-evolution (i.e., the appearance of variations within a species and of species within a genus), which occurs due to the separation and recombination of existing genetic material, macro-evolution requires the production of new genetic material. Such genetic novelty results from a macro-mutational "jump," which in turn produces major morphological changes. I will argue that the macro-evolutionary process is characterized by regulation, direction, and convergence—thus conforming to the Hellenic metaphysical principles of formal and final causality.

As my point of departure, I will outline the metaphysical concept of evolution as the unfolding of inherent possibilities and contrast it with the Neo-Darwinian belief in random mutations upon which natural selection acts in order to produce new forms. I will then consider Presocratic, Platonic, Aristotelian, and Neoplatonic cosmology and bio-philosophy, especially as these pertain to formal and final causality. Specifically, I will discuss the numerical cosmology of Pythagoras and its extension into the organic realm; the teaching of Heraclitus on the *Logos* as cosmic lawgiver and regulator; the creative role of the Demiurge in Plato's cosmology, as well as the interaction between Intellect and Necessity; the fourfold causal scheme of Aristotle, which in the case of living things entails a formation of matter by soul aimed at a specific purpose (*telos*); the Neoplatonic notion of indwelling reason-principles (*logoi*) throughout the manifested order; and the hierarchical scheme of an all-embracing Chain of Being linking the various levels of cosmic reality, including the living kingdoms of plants, animals, and humans, which was an essential element in both the Greek Patristic and Latin Scholastic traditions.

In the main argument of the book, based primarily on the work of the Scottish mathematical biologist D'Arcy Wentworth Thompson, author of the celebrated book *On Growth and Form*, I present an outline of organic form and transformation that emphasizes its mathematical foundations and discuss the limits of transformation. This is followed by a discussion of modern evolutionary theory as understood by the biologists Charles Darwin, Theodosius Dobzhansky, and Stephen Jay Gould, then a critical assessment of the theory from the perspectives of science, philosophy, and metaphysics. In the final three chapters, I argue that the macro-evolutionary process is predominantly regulated, not random, directed, not aimless, and convergent, not divergent, while still allowing for the ubiquitous presence of exceptions—the latter being recognized by both Plato, as Intellect limited by Necessity, and Aristotle, as teleology limited by mechanism.

The conclusion of the book attempts to draw together the domains of the philosophical (*logos*) and the biological (*bios*) in the light of the preceding arguments. It is presented as an invitation to further reflection and debate on this fundamental question, namely the origins and mechanisms of organic diversity on our planet Earth.

1

The Metaphysical Concept of Evolution

IN VIEW OF THE AMBIGUITY of the term "evolution," let us begin by distinguishing between evolution in the traditional, metaphysical understanding and the theory postulated by Charles Darwin and his followers. The term "evolution" is derived from the Latin *evolvere*, literally, to roll out or unfold, which in this context means the development of that which is enveloped.[1] "The cosmos becomes what it is through an 'unwinding' or explication of What is already inside, which is 'turned out' or *evolved* into what It is initially not but can then be seen in."[2] We might say that evolution is the unfolding of inherent possibilities. In Aristotelian terms, we could say that evolution entails a movement from potentiality (Greek *dynamis*) to actuality (Greek *energeia*). Since this unfolding involves an acquisition of form for a specific end or purpose, a *telos*, where intelligible form, *eidos*, serves as the model or template for sensible form, *morphē*, an authentic theory of evolution must include both formal and final causality. This, however, goes against the modern biological understanding of evolution as "the genetic transformation of populations through time, as a result of genetic variation and subsequent environmental impact on the rates of

---

1. Gilson, Etienne, *From Aristotle to Darwin and Back Again. A Journey in Final Causality, Species, and Evolution*, trans. John Lyon (San Francisco: Ignatius Press, 2009), 59.
2. Cutsinger, James S., "On Earth as it is in Heaven," 2007, 11. http://www.cutsinger.net/pdf/ earth_as_it_is_in_heaven.pdf.

reproductive success,"[3] which recognizes only material and mechanical factors as causes of biodiversity. However, in the words of the medieval scholar Richard the Englishman, "Nothing can be produced from a thing that is not contained in it; by this fact, every species, every genus or every natural order develops within the proper limits to it and bears fruits according to its own kind and not according to an essentially different order."[4] That is to say, lifeforms unfold according to pre-determined inner possibilities, although external factors do play a secondary role, as, for instance, in the geographical distribution of a species.

It is popularly held that the idea of evolution results from a modern scientific outlook and effectively replaces previously held theological doctrines on Divine creation, but in reality its origin lies in metaphysics and is grounded in an understanding of reality as a great Chain of Being reaching from God down to inanimate matter. However, "the evolutionary chain of living organisms in post-Darwinian biology is none other than the secularized and temporalized version of the traditional metaphysical doctrine of gradation or the 'great chain of being' of the western tradition."[5]

In his essay *Gradation and Evolution*, Traditionalist writer Ananda Coomaraswamy draws a similar distinction between the traditional, metaphysical doctrine of gradation and the modern, mechanistic theory of evolution. According to the metaphysical view, nothing in the world happens by chance—rather, what happens is always the realization of a possibility, so that all living beings are held to be physical manifestations of inherent possibilities.[6] In this connection, the Swiss metaphysician Frithjof Schuon compares the evolution of the cosmos, and by extension all life in it, to embryonic develop-

---

3. Blackburn, Simon, *The Oxford Dictionary of Philosophy* (Oxford: Oxford University Press, 2008), 123.

4. Quoted in Burckhardt, Titus, "Cosmology and Modern Science," in *The Sword of Gnosis. Metaphysics, Cosmology, Tradition, Symbolism*, ed. Jacob Needleman (Baltimore, MD: Penguin Books, 1974), 147.

5. Bakar, Osman, "The Nature and Extent of Criticism of Evolutionary Theory," in *Science and the Myth of Progress*, ed. Mehrdad Zarandi (2003): 164. www.worldwisdom.com/public/library/default.aspx.

6. Ibid., 166; Coomaraswamy, *Civilization*, 70.

ment: "In the same way the whole cosmos can only spring from an embryonic state which contains the virtuality of all its possible deployment and simply makes manifest on the plane of contingencies an infinitely higher and transcendent prototype."[7] Thus the traditional explanation of the origins of life and all its diversity includes a metaphysical, or vertical, dimension, which determines the physical, or horizontal.

What causes the movement from possibility to manifestation in the organic realm? Coomaraswamy holds that there are two orders of causes: the First Cause, in which all possibilities inhere, and mediate causes, which provide the conditions in which the possible becomes the necessary. The First Cause is the direct cause of the *being* of things, but not of their *manner of being*. The latter is determined by mediate causes that produce a given species or individual at a given time and place. When the mediate causes converge to establish the spatial and temporal environment for a given possibility to be actualized, the corresponding form emerges. This implies, for example, that mammals could not have appeared before the operation of natural causes had prepared the Earth for mammalian life. An evolutionist could validly offer an explanation that employs mediate causes, while omitting the First Cause, because he is dealing with biodiversity and not with the origin of life as such.[8]

The metaphysical distinction between the First Cause and mediate causes is compatible with the theological doctrine of Divine creation, God being the First Cause containing all possibilities, Who creates "indirectly" by means of mediate causes. This theistic understanding of evolution envisions an "evolving creation," as argued by the physicist Howard van Till, building on insights obtained from the Patristic theologians Basil of Caesarea and Augustine of Hippo. Van Till writes that the created order "was gifted from the outset with the potentialities for assuming a rich diversity of physical and biotic forms, and with the requisite form-producing capacities for actualizing many of these potentialities in the course of time," enabling "the unbroken continuity of natural development of the

---

7. Quoted in Cutsinger, *Earth*, 23–24.
8. Coomaraswamy, *Civilization*, 71, 74, 83.

sort envisioned by contemporary cosmology and evolutionary biology."[9] There is no intellectual requirement to posit an incompatibility between theology and biology in the formulation of an evolutionary theory.

In the Patristic understanding, the distinction between ultimate and secondary causes applies also to living beings, and thus "living creatures can originate in two ways: through a primary or 'vertical' mode of generation, which does not involve seed [i.e., procreation] as an intermediate cause; and through a secondary or 'horizontal' mode, that is to say, by means of a natural process. But at the same time we must not forget that the natural process, no less than the primary generation, derives its entire efficacy from the power of God."[10] The distinction between these modes of generation belongs to the realm of manifestation, but the ultimate cause, God, remains the same.

Thomas Aquinas relates the distinction between the First Cause and mediate causes to the universal and the particular orders. The universal order depends on the First Cause of everything and thus embraces everything. In contrast, the particular orders depend on created causes and extend to whatever is subordinate to these causes. He writes, "Necessarily then all these particular orders are subordinate to the one universal order, descendants of that order of things which results in them because of their dependence on the first of causes" (*Summa contra Gentiles*, 3.97–98).[11] However, the distinction between the First Cause and mediate causes should not be conceived as a separation of the physical from the metaphysical. As Schuon emphasizes, "Even within the order of physical causes, one has to take into account the simultaneous presence of the immanent metaphysical Cause: if a seed is the immediate cause of a plant, it is because the divine archetype intervenes in the physical

---

9. Van Till, Howard J., "Basil, Augustine, and the Doctrine of Creation's Functional Integrity," in *Science & Christian Belief* 8:1 (1996): 21–38. http://www.asa3.org/ASA/topics/Evolution/S&CB4-96VanTill.html.

10. Smith, Wolfgang, *Cosmos and Transcendence* (San Rafael, CA: Sophia Perennis, 2008), 89.

11. Aquinas, Thomas, *Selected Philosophical Writings*, trans. Timothy McDermott (Oxford: Oxford University Press, 1993), 275.

The Metaphysical Concept of Evolution

causality."[12] This coherence between the metaphysical and the physical is attributable to their common foundation, both realms having their source in God.

The notion of indirect/secondary yet creative causes also pertains to the concept of natural laws underlying the coming-to-be of the world and of living beings. The Victorian publisher and amateur scientist Robert Chambers held that the formation of the solar system and the Earth, as well as the emergence of life on Earth, are due to the operation of natural laws which are analogous to gravitation in the physical world. These natural laws are established by God, and therefore the formative processes are the work of God acting through secondary causes.[13] In the twentieth century, Christian physicist Howard van Till argues along similar lines with his doctrine of "Creation's functional integrity," which sees creation as "gifted by God from the outset with all of the form-producing capacities necessary for the actualization of the multitude of physical structures and life forms that have appeared in the course of Creation's formative history, and a world whose formational fecundity can be understood only as a manifestation of the Creator's continuous blessing for fruitfulness."[14] These capacities are "sufficiently robust so as not to require additional acts of special creation in time."[15]

According to the traditional understanding of evolution, Coomaraswamy writes, every phenomenon represents one of the possibilities of manifestation of Plato's "ever-productive nature" (*aeigenēs physis*), which is referred to variously in different traditions as God, Spirit, or Life—terms signifying the First Cause of all living beings. God endows all things with life impartially, while leaving the manner of their existence to the operation of mediate causes such as the

---

12. Schuon, Frithjof, *Sophia Perennis and the Theory of Evolution and Progress* (2001): 4. http://www.frithof-schuon.com/evolution-engl.htm.
13. Swift, *Evolution under the Microscope. A Scientific Critique of the Theory of Evolution*, 76. See Robert Chambers' epochal *Vestiges of the Natural History of Creation*, published anonymously in 1844.
14. Van Till, "Basil and Augustine Revisited: The Survival of Functional Integrity," *Origins & Design*, 19:1. http://www.arn.org/docs/odesign/od191/basilaug191.htm.
15. Ibid.

laws of heredity, with which He does not interfere.[16] Although arguing from an atheistic angle, the geologist Paul Lemoine echoes this metaphysical notion of nature's fecundity by postulating an infinite number of living forms that may be dormant but ready to appear when their conditions of existence present themselves.[17]

It is important to note that the metaphysical concept of evolution as the manifestation of inherent possibilities recognizes the observed variability of species, namely that the outward semblance of any genus, species, or individual at any given time or place is always changing, as Coomaraswamy notes.[18] Plutarch reasoned in this regard that "nobody remains one, nor is one; but we become many ... and if he is not the same, we cannot say that he is, but only that he is being transformed as one self comes into being from the other ... and it is only of God, in whose now there is neither future nor past, nor older nor younger, that we can say that He is."[19] Therefore, as Coomaraswamy observes, all definitions of categories like genus, species, and individual are indefinite, since they refer to things that are always becoming. These are only "things" if we ignore their variation in time, which is the relatively short present. Ultimately, every form of life is composite and hence mortal—only beginningless Life, or God, is also endless.[20]

This recognition that the world of phenomena is always subject to becoming repudiates the oft-repeated claim that traditional metaphysics asserts a static world order, which in the organic realm entails fixity of species. It has been suggested, for example, that for the Hellenic philosophers with their teleological world-view, no change in species was believed to be possible, since the design of each species corresponds to an unchanging purpose. Consequently, no new species can arise and there are no gaps in the biosphere to fill.[21] It has also been argued that both Plato and Aristotle viewed the Form of an individual organism as more real than variations

16. Coomaraswamy, *Civilization*, 73, 81, 84.
17. Gilson, *Aristotle*, 153.
18. Coomaraswamy, *Civilization*, 73.
19. Quoted in Coomaraswamy, *Civilization*, 81.
20. Coomaraswamy, *Civilization*, 73–74.
21. Swift, *Evolution*, 59.

The Metaphysical Concept of Evolution

between individuals. Since the Forms are eternal and unchanging, organic phenomena such as species are also conceived as fixed entities.[22] This essentialist perspective is also part of the doctrine of special creation, which teaches that all species of animals and plants that have ever existed were directly created by God, and that the created order requires ongoing direct acts by God in order for new lifeforms to appear in the paleontological record. Howard van Till has rightly criticized this view as presumptuous, since by withholding certain form-producing capacities, God is held responsible for the world being developmentally incomplete.[23]

In contrast to the essentialist position, Plato teaches a dynamic view of the sensible world, depicting the realm of becoming as follows: "It comes to be and passes away, but never really is" (*Timaeus* 28a); and, "What is really true, is this: the things of which we naturally say that they 'are,' are in process of coming to be, as the result of movement and change and blending with one another. We are wrong when we say they 'are,' since nothing ever is, but everything is coming to be" (*Theaetetus* 152d–e). Similarly, for Aristotle all things in the terrestrial realm are subject to alternate generation and decay, in accordance with natural laws (*Parts of Animals* 644b; *On Generation and Corruption* 319a). And in the cosmology of Proclus, universal nature is depicted as entailing three interlinked aspects: the eternally remaining First Principle (*monē*), a procession (*proodos*) thereof through the Forms into beings, and a return (*epistrophē*) of the beings through the Forms unto the First Principle.[24] This Neoplatonic view of procession and return is actually conceived as a single movement. For E.R. Dodds, this diastole-systole is the "life of the universe."[25] In other words, the cosmos is produced through a

---

22. MacNeill, Allen, *The Platonic Roots of Intelligent Design Theory*. www.evolutionlist.blogspot.com/2006/02/platonic-roots-of-intelligent-design.html.

23. Van Till, *Integrity*.

24. I.P. Sheldon-Williams, "The Greek Christian Platonist Tradition from the Cappadocians to Maximus and Eriugena," in *The Cambridge History of Later Greek and Early Medieval Philosophy*, ed. A.H. Armstrong (London: Cambridge University Press, 1967), 431.

25. Quoted in Coomaraswamy, *Civilization*, 83.

ceaseless interaction between being and becoming effected by the alternating motions of procession from the One and return to It.

Following Plato, Coomaraswamy writes that every visible form (*morphē*) of species or individual reflects an archetypical possibility (or Form, *eidos*). Thus there is an invisible Sun, an Apollo, other than the visible sun, or Helios. This traditional doctrine is not monistic or dualistic, but is descriptive of a reality that is both one in itself and many in its manifestations. Accordingly, God is conceived as omniform (*pantomorphos*), while the vast variety of life-forms melt into one another and cannot be precisely defined.[26] That is to say, the cosmos (including the organic realms) comprises a differentiated unity, based in the divine Principle.

That the evolutionary process is none other than an "opening out" of the divine Subject into various degrees of reality is illustrated by the Traditionalist scholar James Cutsinger using the analogy of the point, here representing the infinity of God, and its deployment in space. The point is deployed first from point to line, then from line to plane, and finally from plane to solid, where it is present in a uniform way throughout a particular object, for example in the crystalline structure of a diamond. In order to increase its amplitude, the point then "goes indoors," as it were. This interiorization represents the beginning of organic life as it unfolds on the inside of matter, namely the plant kingdom. At this stage of the evolutionary process the function of specialization appears for the first time, so that the parts become differentiated from the whole. In addition, the more inward character of biological processes provides the divine Subject with additional opportunities for expansion, as is evident in growth. However, although the plant is able to grow and blossom, it is still limited due to its attachment to the Earth. Thus in the next evolutionary stage the point interiorizes further, giving rise to sentience and the power of locomotion, both of which characterize the animal kingdom.[27] This "interiorization" of three-dimensional matter symbolizes the activity first of the

---

26. Coomaraswamy, *Civilization*, 74–75, 79.
27. Cutsinger, *Earth*, 16.

nutritive and then of the sensitive levels of soul, to use Aristotelian terminology.

Since the whole cosmogonic process is grounded in God as ultimate Source, Cutsinger continues, it is the divine Subject deploying Itself through matter as life (i.e., the plant kingdom) and through life as sentience (i.e., the animal kingdom). In this understanding of evolution, in contrast to Darwinian transformism, it is not a question of matter "evolving" into life or life "evolving" into sentience, nor, needless to say, is it an evolution of amphibians into reptiles or of reptiles into birds. Rather, "the only evolution is that of the point, which is the Divine Self as Subject. The forms of existence through which It 'passes,' in a strictly non-temporal and instantaneous way, do not themselves change, for they are the unalterable images of celestial ideas—the distinct and immutable shadows cast by the Divine Sun as It shines upon the eternal archetypes of Its myriad creatures."[28] This notion of the Divine Subject unfolding Itself in and through the evolutionary process was proposed more than two thousand years ago in the profound cosmogony taught by Diogenes of Apollonia: "But all these things (earth, water, air, fire, and all the rest of the things in the cosmos), being differentiated out of the same thing [i.e., God], come to be different things at different times and return into the same thing" (Fragment 2).[29] In other words, the various stages of an evolutionary process in time are none other than the multiple states of a single, non-temporal divine Essence.[30] Accordingly, cosmic reality (including the living kingdoms) is neither monistic nor dualistic, but comprises a differentiated unity, or a many-in-One.

---

28. Ibid., 18.
29. Except as otherwise indicated, extracts from the Presocratic fragments in this work are drawn from McKirahan, Richard, *Philosophy before Socrates. An Introduction with Texts and Commentary* (Indianapolis, IN: Hackett Publishing), 1994.
30. Cutsinger, *Earth*, 13.

2

Aspects of Hellenic Cosmology

WHEN CONSIDERING philosophical and theological terminology, etymology is a useful point of departure. The term "cosmos" is derived from the Greek *kosmos*, which initially meant order, from which it came to signify the world or universe, on account of the latter's (perceived) perfect arrangement.[1] The noun *kosmos* is cognate to the verb *kosmeō*, which means to set in order, or setting things in their proper order. Hence the term *kosmos* also has an aesthetic dimension, in the sense of ornament or adornment.[2] This notion of cosmic setting in order implies the teleological concepts of design and regulation. Related to these latter is the Hellenic notion of harmony (*harmonia*), which originally pertained to a lyre which is in tune. *Harmonia* thus implies balance or being in tune, like the cognate terms *krasis* (mixing, blending) and *synthesis* (putting together, compounding).[3] This etymology suggests that the cosmos is properly ordered and thus displays harmony, both of which can only be brought about by the activity of a transcendent power.

In Hellenic cosmology, we encounter both mechanistic and teleological explanations. Mechanism is the belief that everything can be explained in terms of quantitative laws that govern the interactions of material particles, thereby also explaining the other properties of matter. In biology, mechanism views animals as material

---

1. Liddell, Henry and Scott, Robert eds., *A Greek-English Lexicon* Abridged edition (Oxford: Oxford University Press, 2004), 389; hereafter referred to as LSJ.
2. Vlastos, Gregory, *Plato's Universe* (Oxford: Clarendon Press, 1975), 3.
3. Thompson, D'Arcy Wentworth, *On Growth and Form* Abridged edition (Cambridge and New York: Cambridge University Press, 1992), 7; LSJ, 391, 677.

systems and is therefore hostile towards teleology. In contrast, teleology, from the Greek *telos*, meaning end or ultimate goal, postulates that there are ends or purposes for all things. Teleology is prominent in the Aristotelian view of nature, whence it profoundly influenced Christian theology.[4] The Catholic philosopher Etienne Gilson has pointed out that since both teleology and mechanism belong to the realm of the philosophy of nature, not to that of natural science, the latter cannot "prove" or "disprove" final causality. Moreover, for the philosopher of nature it is not necessary for natural teleology to be perfect in order for us to acknowledge its existence.[5] After all, the cosmos is the product of the interaction between the divine Intellect and irrational Necessity, as Plato says in the *Timaeus*, and it is therefore good but not perfect.

Presocratic Cosmology

Two recurring Presocratic notions that are relevant here are hylozoism and panpsychism. According to the doctrine of hylozoism (from *hylē*, matter and *zōē*, life), all matter is endowed with life. Panpsychism (from *pan*, all and *psychē*, soul), holds that all material things have a level of consciousness. Since the cosmos produces living creatures, it should itself be thought of as alive and possessing a world soul.[6] The ecological implications of hylozoism and panpsychism are manifold, not the least of which is a damning indictment of the ever-increasing human exploitation and destruction of the Earth and its biosphere.

According to Anaximander of Miletus (c. 610–547 BC), there exists a cosmic justice (*dikē*) that maintains the balance among the primary elements of water, earth, air and fire. This notion is implied by a fragment preserved by the Neoplatonist commentator Simplicius (fl. 6th c. AD): "The things that are perish into the things out of which they come to be, according to necessity, for they pay penalty and retribution to each other for their injustice in accordance with the ordering of time, as he [Anaximander] says in rather poetical

---

4. Blackburn, *Philosophy*, 228, 360.
5. Gilson, *Aristotle*, 20, 143.
6. Blackburn, *Philosophy*, 174, 265.

language."[7] By "paying penalty and retribution to each other," domination of one element over the others is prevented, which in turn preserves the structural integrity of the cosmos throughout the ages.[8] Anaximander thus became the first Western thinker to conceive of the cosmos as ruled by lawful necessity.[9] As we read in the Pythagorean *Golden Verses*, an influential collection of moral teachings, "And thou shalt know that law ... established the inner nature of all things alike."[10] This concept of a lawfully ordered cosmos has been of lasting significance, both to metaphysics and natural science.

One of the most influential Hellenic philosophers, Pythagoras of Samos (c. 582–496 BC), has traditionally been honored as the father of Western mathematics. According to the early Christian historian Eusebius, Pythagoras invented the term "philosophy," wishing to be called a lover of wisdom, or *philosophos*.[11] The point of departure for Pythagorean philosophy was the discovery that concordant musical intervals can be expressed mathematically. To be more precise, "the first natural law ever formulated mathematically was the relationship between musical pitch and the length of a vibrating harp string."[12] Pythagoras and his students noticed that certain ratios of string lengths always produce the harmonic intervals of the octave (2:1), the fifth (3:2), and the fourth (4:3). The sum of these four numbers (1+2+3+4) provides the sacred number ten, the *tetraktys*, meaning "fourness." This scheme is geometrically represented by four rows of pebbles arranged from four at the base to

---

7. Quoted in McKirahan, *Philosophy*, 43.

8. Theodossiou, Efstratios, Manimanis, Vassilios N. & Dimitrijevic, Milan S., "The cosmological theories of the pre-Socratic Greek philosophers and their philosophical views for the environment," in *Facta Universitatis*, 10:1 (2011): 94.

9. Dreyer, P.S., *Die Wysbegeerte van die Grieke* [The Philosophy of the Greeks], Kaapstad & Pretoria: HAUM (1975): 31.

10. Critchlow, Keith, Foreword to *Quadrivium. The four classical liberal arts of number, geometry, music & cosmology*, ed. John Martineau (Glastonbury: Wooden Books, 2010), 5.

11. Bailey, Jim, *Sailing to Paradise, The Discovery of the Americas by 7000 BC* (New York: Simon & Schuster, 1994), 274.

12. Ferguson, Kitty, *Pythagoras. His Lives and the Legacy of a Rational Universe* (London: Icon Books, 2011), 62.

Aspects of Hellenic Cosmology

one at the summit, thus forming an equilateral triangle consisting of ten pebbles. The followers of Pythagoras also discovered that a tetrahedron, or four-sided solid, could be constructed out of four equilateral triangles. The focus on form provided physical natures with an intelligible grounding in different geometrical structures,[13] and anticipated the insistence by both Plato and Aristotle on the priority of form over matter in the constitution of physical reality.

In the Pythagorean cosmology, all things arise from the one (*monas*), which is the number of the Godhead. Number is therefore the cosmic principle from which all things arise. At first a relation is established by the Pythagoreans between numbers and geometry, with the basic entity in these realms being the unit (*monas*), i.e., a point lacking position, and the point (*stigmē*), i.e., a unit having position, respectively. Numbers are pluralities of units, while lines, planes, and solids are extensions of points into one, two, and three dimensions.[14] Furthermore, the Pythagoreans held that each of the four basic elements is derived from a particular geometrical figure: earth is made from the cube, fire from the pyramid, air from the octahedron and water from the icosahedron, while the sphere of the whole is made from the dodecahedron. These shapes would have been familiar to the early Pythagoreans from both natural and artificial constructions: cubes and pyramids were often used in building, and various kinds of crystal appear as cubes, octahedra or dodecahedra. In this way the physical nature of the universe is accounted for in terms of basic kinds of matter, while matter itself is analyzed in terms of a finite number of simple geometrical bodies.[15]

Accordingly, for the Pythagoreans the progression from the transcendent Principle to the realm of immanent manifestation entails a movement from numbers into the physical world by means of geometrical figures. Interestingly, although the cosmos is numerically grounded, the realm of space in turn imposes limits on number. Thus among perfect polygons one encounters three regular grids, five regular solids (also known as the Platonic solids, i.e., the tetra-

---

13. Ibid., 63, 69; Blackburn, *Philosophy*, 300.
14. Dreyer, *Wysbegeerte*, 36; McKirahan, *Philosophy*, 100.
15. Ferguson, *Pythagoras*, 155–56; McKirahan, *Philosophy*, 102.

hedron, cube, octahedron, dodecahedron, and icosahedron), eight semi-regular grids, and thirteen semi-regular solids.[16] It is immediately evident that these four numbers (3, 5, 8 and 13) are interrelated: three plus five gives eight, and five plus eight gives thirteen. In its turn thirteen is the number of semi-regular polyhedra, also known as Archimedean solids.[17] We will encounter this particular instance of mathematical regulation again in a later chapter dealing with the numerical grounding of organic forms.

In the realm of ontology, Pythagoras and his followers realized the importance of the quantitative aspect of being. Pythagorean numerology implies a distinction between the formal and the material, and is the first appearance of this fundamental insight in Hellenic thought. Matter as such is held to be nothing; only when bounded does it assume form and become *something* (since form is the expression of limitation). Being is thus made possible through the unity of matter and form.[18] Philolaus states in this regard that the fundamental principles of the limitless (*apeiron*) and the limiting (*peras*) have to be "locked together" by *harmonia* in order to produce the world of differentiation.[19] This Pythagorean understanding of the cosmos being the result of the interaction between form and matter would be thoroughly elaborated by Plato, Aristotle, and the Neoplatonists.

The enigmatic Heraclitus of Ephesus (c. 535–475 BC) achieved lasting philosophical and theological relevance with his dynamic conception of physical reality and his doctrine of logos. Of the logos, he writes: "This *logos* holds always but humans always prove unable to understand it, both before hearing it and when they have first heard it. For though all things come to be in accordance with this *logos*, humans are like the inexperienced when they experience such words and deeds as I set out, distinguishing each in accordance with its nature and saying how it is" (Fragment 1); and, "For this reason it is necessary to follow what is common. But although the

---

16. Lundy, Miranda, "Sacred Number," in *Quadrivium*, 38.
17. Sutton, Daud, "Platonic and Archimedean Solids," in *Quadrivium*, 162.
18. Dreyer, *Wysbegeerte*, 35, 38.
19. Ferguson, *Pythagoras*, 106–07.

logos is common, most people live as if they had their own private understanding" (Fragment 2).[20] The noun *logos*, related to the verb *legō*, meaning to relate, speak, or say, means the word by which the inward thought is expressed and also the inward thought or reason itself. From this basis, further meanings of the words *word*, *story*, or *reason* are derived.[21] Martin Heidegger remarked that from Fragments 1 and 2 we can deduce Heraclitus' understanding of the *logos*, namely that it is constant; that it unfolds as the "together" in beings; and that everything which happens is in accordance with this constant "together." In other words, Heidegger writes, *logos* means the constant gathering of beings that stands in itself.[22] Moreover, due to the presence of the *logos*, reality displays both unity and plurality. In the words of Heraclitus, "out of all things there comes a unity, and out of a unity all things" (Fragment 48). The activity of the *logos* guarantees that all things are one and one thing is all—an insight which Heraclitus presents in terms of the interaction between a variety of opposites.[23] Evidently, for Heraclitus, the *logos* is the link between the One and the many; that is to say, between cosmic unity and diversity.

Heraclitus has often been credited with teaching that all existing things are in a state of flux. For instance, in Plato's epistemological dialogue, *Theaetetus*, Socrates mentions a theory held by Heraclitus that all things are in flux and motion, so that nothing is stationary (152e, 156a, 179e). This notion is expressed in Greek as *panta rei*: "everything flows." One cannot step into the same river twice, Heraclitus argued, for different waters are always flowing in it (Fragments 12 and 91). Therefore, in the phenomenal world nothing *is*, but everything *becomes*.[24] It has even been suggested that Heraclitus anticipated quantum physics with his teaching that the apparent stability of the natural world is an illusion of our senses.[25] However,

---

20. McKirahan, *Philosophy*, 116.
21. LSJ, 416.
22. Heidegger, Martin, *Introduction to Metaphysics*, trans. Gregory Fried & Richard Polt (New Haven: Yale University Press, 2000), 135, 138.
23. McKirahan, *Philosophy*, 134–35.
24. Dreyer, *Wysbegeerte*, 41.
25. Theodossiou et al., *Theories*, 90.

not all commentators accept this "conventional" view of Heraclitus. For example, Richard McKirahan writes that the Ionian philosopher appears to value equally change and stability, plurality and unity, difference and identity, one pole necessitating the existence of the other. This reading is confirmed by Fragment 8, on the harmony of opposites: "What is opposed brings together; the finest harmony is composed of things at variance, and everything comes to be in accordance with strife." As a matter of fact, without differentiation there would be no harmony, since harmony is a relation among different things.[26]

Heraclitus also holds that the cosmos consists of pairs of opposites that are continuously interacting. This interaction has a theistic dimension, as suggested in Fragment 67, where we are told that God is related to various pairs of opposites: day and night, winter and summer, and even war and peace. The *logos* rules the cosmos by regulating the conflict between opposites and thereby provides a deeper unity and harmony that underlie the changes flowing from their conflict. For Heraclitus, all things find their unity in the *logos*, while the *logos* manifests itself in the multiplicity of phenomena.[27] As cosmic regulator, the *logos* is the mediator between the One and the many. In addition, Heraclitus holds that the order of the cosmos is not only rational, and thus in principle intelligible, but "it is also an intelligent system: there is an intelligent plan at work, if only in the sense of the cosmos working itself out in accordance with rational principles."[28] It is the *logos* that is responsible for the rationality of the cosmos.

The first known philosopher to teach in Athens, thenceforth the intellectual center of Hellenic civilization, was Anaxagoras of Clazomenae (c. 500–428 BC). In his writings he depicted an original cosmic state in which all things were mixed. Then an external force, the *nous*, translated as mind or thought, acted within the primeval mass of *spermata* by causing a rotation to begin. The rule of *nous* or

---

26. McKirahan, *Philosophy*, 134, 142.
27. Dreyer, *Wysbegeerte*, 42.
28. Curd, Patricia, *Presocratic Philosophy*, Stanford Encyclopedia of Philosophy (2011). http://plato.stanford.edu/entries/presocratics.

Aspects of Hellenic Cosmology

Mind over all things results from its omnipresence and its ability to cause motion. Motion, in its turn, causes all changes in the phenomenal world by means of mixture (*synkrisis*) and separation (*diakrisis*). According to Anaxagoras, "The Greeks are wrong to accept coming to be and perishing, for no thing comes to be, nor does it perish, but they are mixed together from things that are and they are separated apart. And so they would be correct to call coming to be being mixed together, and perishing being separated apart" (Fragment 17). Thus, all the motions of material things can ultimately be traced to the action of Mind. Mind rules over all things through its power of causing them to move, not in a random fashion but in a way that sets them in order. But, for Anaxagoras, the action of Mind in establishing the cosmos is not of a purely mechanical nature, but also has an intellectual dimension. As stated in Fragment 12: "And Mind rules all things that possess life—both the larger and the smaller. . . . And Mind knew all the things that are being mixed together and separated off and separated apart. And Mind set in order all things, whatever kinds of things were to be—whatever were and all that are now and whatever will be."[29] Evidently, since Mind rules all things, knows all things, and orders all things, the philosophy of Anaxagoras entails a teleological rather than a mechanistic world-view.

Anaxagoras' contemporary, Empedocles of Acragas (c. 495–435 BC), introduced his treatise *On Nature* with a depiction of four basic substances (or elements) and two sources of change. The four elements, which are eternal and ungenerated, are earth, water, air, and fire. These elements are also called *rhizōmata* (plural of *rhizōma*, meaning root), and according to Empedocles all existing things consist of quantitative mixtures thereof. He therefore agrees with Anaxagoras that the origin and cessation of things is a process of mixture and separation, which is never-ending. In addition to the primary elements there are two sources of change: Love (*Philia*) and Strife (*Nikos*). While Love unites and mixes unlike things, Strife sets unlike things in opposition and instead mixes like with like. The ever-changing cosmos is therefore the result of intermediate

---

29. LSJ, 467; McKirahan, *Philosophy*, 198–200, 203, 219, 223.

phases between the extremes produced by the triumph of either Love or Strife.[30]

The last of the Presocratic cosmologists was Diogenes of Apollonia (5th c. BC), a younger contemporary of Socrates. Diogenes held that the entire world of physical phenomena arises from the intelligence (*noēsis*) underlying it. The term *noēsis* is related to *nous*, which as we have noted was used by Anaxagoras to denote Mind. As we read in Fragment 5 of Diogenes' writings, "That which possesses intelligence is what people call air, and all humans are governed by it and it rules all things. For in my opinion this very thing is god, and it reaches everything and arranges all things and is in everything. And there is no single thing which does not share in this." Air, which is divinity and intelligence and rule, is the fundamental principle of things. Diogenes continues, "And this very thing is an eternal and immortal body, and by means of it some things come to be and others pass away" (Fragment 7). Moreover, as Richard McKirahan comments, Diogenes anticipated the celebrated Argument from Design by holding a teleological world-view. In Fragment 3 we read, "For without intelligence it could not be distributed in such a way as to have the measures of all things. . . . If anyone wants to think about the other things too, he would find that as they are arranged, they are as good as possible." For Diogenes, the order in the universe is conceived as the result of intelligence, since if everything is arranged in the best possible way, it follows that the cause of that arrangement is intelligent.[31] That is to say, the cosmos is characterized by final causality due to its intelligent arrangement, or design.

The Cosmology of Plato

The most celebrated ontological concept of Plato (c. 429–347 BC) is that of the transcendent Forms or Ideas (Greek *ideai*) that serve as models, or templates, for the world of phenomena. The wide-ranging scope of the Forms has been accurately summarized by Desmond Lee in his translation of the *Republic*:

---

30. LSJ, 625; Dreyer, *Wysbegeerte*, 51; McKirahan, *Philosophy*, 259; Curd, *Presocratic Philosophy*.
31. McKirahan, *Philosophy*, 345–48.

> Briefly therefore we may say that the forms are objects of knowledge (as opposed to opinion), are what is ultimately real (as opposed to what appears or seems), are standards or patterns to which different but similar particulars approximate, though imperfectly (this meaning is particularly relevant in morals and mathematics), and are the common factor in virtue of which we give groups of particular things a common name.[32]

In other words, the Forms have epistemological, ontological, paradigmatic, and nomenclatural dimensions.

The Forms are part of the intelligible world, which Plato distinguishes from the sensible world:

> As I see it, then, we must begin by making the following distinction: What is that which always is and has no becoming, and what is that which always becomes but never is? The former is grasped by understanding (*noēsis*), which involves a reasoned account (*logos*). It is unchanging. The latter is grasped by opinion (*doxa*), which involves unreasoning sense perception (*aisthēsis alogos*). It comes to be and passes away, but never really is. (*Timaeus* 27d–28a)

In this passage, Plato affirms both the realm of becoming, as did Heraclitus, and the realm of being, as did Parmenides.[33] It is important to note that this notion of an eternal, intelligible realm that is reflected in the temporal, sensible realm does not imply a cosmological dualism, since these realms imply each other. As Coomaraswamy writes, "The uniformity of the intelligible world is in every way compatible with the multiformity of its manifestations."[34]

In his late dialogue, the *Timaeus*, Plato presents a cosmology that would achieve lasting significance in Western philosophy and theology. The *Timaeus* introduces for the first time in Hellenic philosophy a comprehensive scheme of creation by a divine Craftsman, so that the world resembles a work of art that is designed with a purpose, the Forms serving as models or patterns (*paradeigmata*) on

---

32. Lee, Desmond, trans., *The Republic* (London: Penguin Books, 1987), 266.
33. Dreyer, *Wysbegeerte*, 94.
34. Coomaraswamy, *Civilization*, 72.

which the maker of the cosmos based all things.³⁵ Plato calls the world-maker the Father, or more often, the Demiurge (*ho Demiourgos*). Significantly, the goodness of the Demiurge is stated as the motive for creation:

> He was good, and one who is good can never become jealous of anything. And so, being free of jealousy, he wanted everything to become as much like himself as was possible.... The god wanted everything to be good and nothing to be bad so far as that was possible, and so he took over all that was visible—not at rest but in discordant and disorderly motion—and brought it from a state of disorder to one of order, because he believed that order was in every way better than disorder.... Accordingly, the god reasoned and concluded that in the realm of things naturally visible no unintelligent thing could as a whole be better than anything which does possess intelligence as a whole, and he further concluded that it is impossible for anything to come to possess intelligence apart from soul. Guided by this reasoning, he put intelligence in soul, and soul in body, and so he constructed the universe.... This, then, in keeping with our likely account, is how we must say divine providence brought our world into being as a truly living thing (*zoion*), endowed with soul and intelligence. (29e–30c)

Again, the transformation of pre-cosmic chaos into cosmic order entails a process of implanting intelligence into soul, and soul (which is life, since *psychē* is translated as breath, life, or soul)³⁶ into body or the physical aspect of being.

Plato acknowledges that order is not inherent in the sensible world, but needs to be imposed by Intellect as personified by the Demiurge. Regarding the ontological status of Intellect, "it is reasonable to conclude that Intellect is a *sui generis* substance that transcends the metaphysical dichotomy of being and becoming—possibly not unlike the Judeo-Christian conception of God."³⁷ Here

---

35. Dreyer, *Wysbegeerte*, 94; Cornford, Francis, *Plato's Cosmology. The Timaeus of Plato* (Indianapolis, IN: Hackett Publishing, 1997), 31.
36. LSJ, 798.
37. Zeyl, Donald, *Plato's Timaeus*, Stanford Encyclopedia of Philosophy (2009). http://plato.stanford.edu/entries/plato-timaeus.

Aspects of Hellenic Cosmology

it is worth noting that the literal meaning of *hylē* (matter) is wood, forest, or woodland, as opposed to *dendra*, meaning fruit trees. Generally speaking, *hylē* refers to the stuff of which a thing is made, the raw material. It is thus fitting that the transcendent power by which all things are made is depicted as an architect or a carpenter.[38] The Platonic doctrine of a Divine Craftsman reflects this notion of a transcendent power shaping raw materials. However, the Demiurge is not omnipotent in its creative work, but is constrained by Necessity (*anangkē*, also translated as force or constraint). In the *Timaeus*, Necessity indicates the indeterminate, the inconstant, and the anomalous, a force that is irregular and unintelligible. Although irrational, Necessity resides in the properties of the elements: fire, for example, has the characteristic power (*dynamis*) to produce burning heat. By affirming the role of Necessity in the constitution of the cosmos, Plato admits that the cosmos contains active powers that are independent of the divine Intellect and are always producing undesirable effects.[39]

The physical cosmos is presented by Plato as the offspring of the union of Intellect and Necessity: "For the generation of this universe was a mixed result of the combination of Necessity and Intellect. Intellect overruled Necessity by persuading her to guide the greatest part of the things that become towards what is best; in that way and on that principle this universe was fashioned in the beginning by the victory of reasonable persuasion over Necessity" (*Timaeus* 48a). "That is why we must distinguish two forms of cause, the divine and the necessary" (68e) in the production of the cosmos. It is our contention that Plato's conception of the physical world as the result of co-operation between Intellect and Necessity represents an advance on the Judaic-Christian cosmology according to the book of Genesis ("And God saw that it was all good"), since Nature displays both splendid design and purpose on the one hand and rampant suffering and waste on the other.

Before describing the nature of the receptacle of becoming (*hypodochē*), an undifferentiated material substratum that underlies

---

38. LSJ, 725; Coomaraswamy, *Civilization*, 82.
39. LSJ, 58; Cornford, *Timaeus*, 171–72, 174, 176.

the sensible world,[40] Plato shows how the four elements continually change into each other. For instance, water becomes air by a process of condensation. Since the elements possess no stability, everything which is composite is unstable. It is therefore better to say "what is such" than "this or that" to describe a thing (*Timaeus* 49c–50a). The Athenian philosopher thus finds himself in agreement with Heraclitus that nothing in the realm of becoming is substantial. Plato precedes his account of the formation of the physical universe with a vivid description of the initial chaos out of which the Demiurge creates the cosmos (*Timaeus* 52d–53c). The task of the divine Craftsman is to transform the inchoate primordial matter from chaos into cosmos by imposing form onto it.[41] That is to say, creation is not out of nothing (*ex ouk onton*) as in the Christian tradition, but out of formless matter (*ex amorphou hylēs*).

Initially the four "kinds" (*genē*; Plato prefers this term to "elements," *stoicheia*) of fire, air, water, and earth are present in the receptacle, but without proportion and measure. They are "thoroughly god-forsaken" in their natural condition, and therefore the Demiurge has to give these kinds their distinctive shapes, by means of form and numbers (*Timaeus* 53a–b). The initial creative activity of the Demiurge entails the fashioning of the elements according to specific geometrical figures known as regular solids. He chooses the tetrahedron for fire, the octahedron for air, the icosahedron for water, and the cube for earth (55e–56a). Through the imposition of order onto the pre-existent chaos of the universe, the Demiurge establishes intelligibility in the cosmos. And since this cosmic ordering is brought about by means of mathematical shapes and numbers, for Plato all that is intelligible in the sensible world is mathematically expressible. However, this should not be understood to mean that the sensible world contains only what is mathematically expressible.[42] Instead, we contend, through the interaction between

---

40. Zeyl, 2009.
41. Vlastos, *Universe*, 70.
42. Gerson, Lloyd P., *Aristotle and Other Platonists* (Ithaca, NY: Cornell University Press, 2005), 215, 219, 240.

Aspects of Hellenic Cosmology

Intellect and Necessity, both final and mechanistic causation are established in the cosmos.

Plato concludes the *Timaeus* with a eulogy to the created order:

> And so now we may say that our account of the universe has reached its conclusion. This world of ours has received and teems with living things, mortal and immortal. A visible living thing containing visible ones, perceptible god, image of the intelligible Living Thing, its grandness, goodness, beauty and perfection are unexcelled. Our one universe, indeed the only one of its kind, has come to be. (92c)

However, Plato cautions us, because the physical universe is perceptible, we cannot have true knowledge (*epistēmē*) of it, but only an opinion (*doxa*). He therefore emphasizes that the *Timaeus* is a "likely account" of the cosmos and its creation (29d, 72d). According to Plotinus, this statement affirms that the anthropomorphisms ascribed to the Demiurge such as sowing or speaking, are not to be taken literally, particularly where these actions imply temporality (*Enneads* IV, 8, 4).[43] This argument represents a remarkable objection in the late Hellenic era to what in recent times has become known as literal creationism.

The Metaphysics of Aristotle

An eminent student of Plato for around twenty years, Aristotle (384–322 BC) divides the sciences into three distinct types: theoretical, practical, and productive, of which only the first is relevant to our argument. In turn, the theoretical sciences are subdivided into theology or metaphysics, physics, and mathematics, respectively dealing with substances unconnected with matter, natural bodies, and numbers and spatial figures (*Metaphysics* VI.1025b–1026a).[44] That mathematics should not be limited to its quantitative dimension was self-evident to Aristotle, who wrote that "the mathematical sciences particularly exhibit order, symmetry, and limitations; and

---

43. Dillon, John and Gerson, Lloyd, eds., *Neoplatonic Philosophy. Introductory Readings* (Indianapolis, IN: Hackett Publishing, 2004), 62.

44. Ross, David, *Aristotle* (London and New York: Routledge, 1995), 65.

these are the greatest forms of the beautiful."[45] Interestingly, symmetry literally means "measuring together" (from the Greek *symmetreia*), so that the underlying relation with geometry and harmony immediately becomes apparent.[46]

Aristotle viewed theology as the first and highest science, since it "deals with things which both exist separately and are immovable" (*Metaphysics* VI.1026a)—that is, the realm of the Divine. Contrary to the reigning paradigm in the modern natural sciences, Aristotle refused to divorce the immanent from the transcendent, or the physical from the metaphysical. This is evident from the opening sentence of *On the Heavens*, where he declares that the science of nature concerns itself with bodies and magnitudes, their properties and movements, and their principles. He adds: "For of things constituted by nature some are bodies and magnitudes, some possess body and magnitude, and some are principles of things which possess these" (I.268a). Therefore, Aristotle writes in the same section, natural science deals not only with the elements, but also with animate things such as plants and animals, and their principles such as matter, form, movement, and soul.

While Plato distinguished between the sensible world of becoming and the intelligible world of being, Aristotle made a distinction between the terrestrial and celestial worlds, respectively. As stated in the *Parts of Animals*, "Of things constituted by nature some are ungenerated, imperishable, and eternal, while others are subject to generation and decay. The former are excellent beyond compare and divine, but less accessible to knowledge" (I.644b). In other words, for Aristotle the terrestrial world is imperfect and mutable, consists of the four elements, and is subject to natural laws, whereas the celestial world is perfect and immutable, consists of ether, and is itself the realm of supernatural laws.[47] Since only the sub-lunar world is ruled by natural laws, it is for Aristotle the domain of natural science.

---

45. Quoted in Marshall, Perry, *Evolution 2.0: Breaking the Deadlock between Darwin and Design* (Dallas, TX: BenBella Books, 2015), 214.
46. LSJ, 664; Lundy, Miranda, "Sacred Geometry," in *Quadrivium*, 104.
47. Swift, *Evolution*, 19.

Aspects of Hellenic Cosmology

Aristotle conceives of substance (*ousia*; also rendered the being or essence of a thing) as the primary category of reality. The scope of the term "substance" is stated thus in the *Metaphysics*:

> Substance is thought to belong most obviously to bodies; and so we say that not only animals and plants and their parts are substances, but also natural bodies such as fire and water and earth and everything of the sort, and all things that are either parts of these or composed of these (either of parts or of the whole bodies), e.g., the physical universe and its parts, stars and moon and the sun. (VII.1028b)

In other words, all things that exist in the physical sense are viewed as substances.

Aristotle's notion of substance may be analyzed from two perspectives: form and matter, and potency and actuality. From a static perspective, each substance consists of form (*eidos*) and matter (*hylē*). In the *Metaphysics*, Aristotle illustrates his argument by means of a bronze statue: the matter is the bronze, the form is the shape or pattern, and the concrete whole, the substance, is the statue. The form is prior both to the matter and the compound (VII.1029a). Aristotle employs both of the terms *morphē* and *eidos* to signify form, but generally uses *morphē* for sensible shape and *eidos* for intelligible structure. Evidently for Aristotle the term *eidos* indicates the Platonic Form, albeit always viewed in conjunction with matter. As such, form determines the essence (*ousia*) of a thing. Accordingly, matter is a relative term, and to each form there corresponds a special matter (*Physics* II.194b). Thus, in living beings, the elements are matter relative to their simple compounds, namely tissues; the tissues are matter relative to the organs; and the organs are matter relative to the living body.[48]

From a dynamic perspective, each substance consists of potency and actuality. The terms potency and potentiality are derived from the Latin *posse*, "to be able," from which also derives *possibilis* or the possible. Accordingly, to be potent means to be rich in possibili-

---

48. LSJ, 507; Dreyer, *Wysbegeerte*, 130–31; Ross, *Aristotle*, 76.

ties.⁴⁹ The Greek term *dynamis* primarily means power, and in the *Metaphysics* Aristotle distinguishes two senses thereof: the power to produce change in something else (VIII.1046a); and the potentiality in a thing to pass from one state into another, as the wood might be said to have the potentiality to become a statue (IX.1048a). Aristotle associates matter with potency and form with actuality, and just as form is prior to matter, actuality is prior to potency. In the *Metaphysics* he writes: "Further, matter exists in a potential state, just because it may come to its form; and when it exists actually, then it is its form" (IX.1050a). Thus, to actualize a possibility is to give form to matter. A particular matter only contains certain possibilities, and therefore matter depends on form for its realization. For example, stones and wood contain the possibility of being used to build a house.⁵⁰

Aristotle held that all existing things come into being through the interaction between four kinds of causes. The material cause is the substance or material that constitutes something; the formal cause is the pattern or blueprint determining the form of the result; the efficient cause is the agency producing the result; and the final cause is the end towards which the result is produced.⁵¹ For Aristotle, each of the four causes could be the proximate cause or the distant cause, and each cause is viewed as a condition that is necessary but not separately sufficient to account for the existence of a thing. In general, all four causes are required to produce any effect.⁵² From a Platonic viewpoint, the material cause is of an immanent nature (denoting the receptacle of becoming), while the formal cause is transcendent (being one or more of the eternal Forms). The final cause "draws together," as it were, the material, formal and efficient causes in order to produce the intended result.

Aristotle also examines the roles of chance and spontaneity in so

---

49. Schuon, Frithjof, *From the Divine to the Human. Survey of Metaphysics and Epistemology*, trans. Gustavo Polit & Deborah Lambert (Bloomington, IN: World Wisdom Books, 1982), 43.
50. Dreyer, *Wysbegeerte*, 133.
51. Blackburn, *Philosophy*, 57.
52. Ross, *Aristotle*, 75. See also *Physics* II.194b; *Metaphysics* V.1013a.

far as they act as causes of things coming to be. In the *Physics*, he writes: "It is clear then that chance is an incidental cause in the sphere of those actions for the sake of something which involve purpose. Intelligent reflection, then, and chance are in the same sphere, for purpose implies intelligent reflection." Further: "Both are then, as I have said, incidental causes—both chance and spontaneity—in the sphere of things which are capable of coming to pass not necessarily, nor normally, and with reference to such of these as might come to pass *for the sake of something*" (*Physics* II.197a; italics ours). Aristotle reaffirms his teleological point of view by concluding, "Spontaneity and chance, therefore, are posterior to intelligence and nature" (II.198a). This implies that even chance events are not anti-teleological in nature, but rather act as incidental causes in the attainment of purpose.

However, Aristotle also recognizes that there are cases in nature where mechanism alone is at work, without the operation of final causality. This is due to the hypothetical necessity of matter, as stated in the *Physics*: "What is necessary then exists by hypothesis and not as an end (*telos*); for it exists in matter, while the final cause is in the account (*logos*)" (II, 200a).[53] Aristotle adds that in this context, "there are then two causes, namely, necessity and the final end. For many things are produced, simply as the results of necessity" (*Parts of Animals* I.642a). Therefore certain phenomena are to be explained only by material and efficient causes, for example the color of a person's eyes. Sometimes necessity even opposes teleology, as in the case of monstrous births that are due to defective matter (*Generation of Animals* 778a–b, 767b).[54]

Among Aristotle's extant writings, Book XII of the *Metaphysics* contains the clearest exposition of his theology. In this treatise, the Prime Mover is defined as the first principle of movement whose essence is actuality (XII.1071b). The nature of the Prime Mover is outlined as follows:

---

53. Translation by Gerson, *Aristotle*, 122.
54. Ross, *Aristotle*, 82.

> And life (*zōē*) also belongs to God; for the actuality of thought is life, and God is that actuality; and God's self-dependent actuality is life most good and eternal. We say therefore that God is a living being, eternal, most good, so that life and duration continuous and eternal belong to God; for this *is* God. (XII.1072b)

Aristotle also provides an ontological dimension to the role of God as Prime Mover. In *On Generation and Corruption* he argues as follows:

> Now "being"... is better than "not-being": but not all things can possess being, since they are too far removed from the "originative source." God therefore adopted the remaining alternative, and fulfilled the perfection of the universe by making coming-to-be uninterrupted: for the greatest possible coherence would thus be secured to existence, because that "coming-to-be should itself come-to-be perpetually" is the closest approximation to eternal being. (II.336b)

In other words, for Aristotle ceaseless becoming is an overcoming of non-being and a striving towards the divine.

The Platonic doctrine that the world is generated yet eternal (since the Demiurge orders formless matter rather than creating it from nothing), is criticized by Aristotle on the grounds that generated things are always observed to be destroyed, while an eternal state is impervious to change. That which always exists is imperishable and ungenerated, Aristotle reasons, since it is incapable of alternating between being and non-being (*On the Heavens* I.279b-280a). Nonetheless, Aristotle concurs with the statement in the *Timaeus* (30a) that the ordered arose out of the unordered, arguing that since the same thing cannot be simultaneously ordered and unordered, a process and a time span is required to separate the two states (*On the Heavens* I.281b). Aristotle also agrees with Plato on the priority of the eternal (*to aionion*) to the transitory, writing: "For eternal things are prior in substance to perishable things, and no eternal thing exists potentially" (*Metaphysics* IX.1050b).[55]

---

55. Gerson, *Aristotle*, 193.

Neoplatonic Cosmology

In a collection of treatises titled the *Enneads*, Plotinus (c. 205–270) laid the cosmological and metaphysical foundations of Neoplatonism. Plotinus viewed all of reality as grounded in the radically transcendent One (*to hen*), from which all beings arise through a process of emanation. In the words of Plotinus: "The One is perfect because it seeks for nothing, and possesses nothing, and has need of nothing; and being perfect, it overflows, and thus its super-abundance produces an Other" (*Enneads* V.2.1).[56] Or, as vividly depicted by the philosopher Michael Grant, the One "pours itself out in an eternal downward rush of generation which brings into being all the different, ordered levels of the world as we know it, in a majestic, spontaneous surge of living forms."[57] The One precedes all things and is simultaneously immanent in all things, establishing ontological continuity throughout the cosmos.

After the One, Plotinus distinguishes between three divine hypostases (*hypostaseis*), or modes of being: "There is the One beyond Being; ... next, there is Being and Intellect; and third, there is the nature of the Soul" (*Enneads* V.1.10). The second hypostasis of the Divinity is the Intellect (*ho nous*), which emanates from the One through its self-contemplation. The contemplation of the One by the Intellect gives rise to an immense variety of separate thoughts, or intelligibles (*noeta*), which are reflections of the power (*dynamis*) of the One that brought the Intellect into existence. These intelligibles are none other than the eternal Forms as postulated by Plato, and through them the Intellect is present in all things as their being and intelligibility.[58] Plotinus employs the Stoic term *logoi spermatikoi*, or seminal reasons, to indicate the productive "seeds" that become actualized as distinct from the Intellect (*Enneads* V.9.6–7). These "rational seeds" contain the potentialities of all beings.[59] The

---

56. Quoted in Lovejoy, Arthur, *The Great Chain of Being. A Study of the History of an Idea* (New York: Harper & Brothers, 1960), 62.
57. Quoted in Ferguson, *Pythagoras*, 200.
58. Moore, Edward, *Neoplatonism*, in Internet Encyclopaedia of Philosophy (2005). http://iep.utm.edu/neoplato.
59. Moore, Edward, *Plotinus*, in Internet Encyclopedia of Philosophy (2001). http://www.iep.utm.edu/plotinus.

third hypostasis of the Divinity is the Soul (*he psychē*), which emanates from the Intellect through its contemplation of the One. In its turn, the Soul contemplates the Intellect, and this causes the emanation of the cosmos. As generator of and ruler over the material world, the Soul forms material beings according to their prototypes in the Intellect (i.e., the Forms of Plato). Accordingly, for Plotinus the World-soul (*psychē tou pantos*) contains the realm of nature (*physis*) as an emanation from the One via the Intellect. Nature is in fact the level where the Soul becomes fragmented into individual, embodied souls.[60] And since the position of the individual soul is intermediate between the World-soul and nature, it displays points of congruence with both realms.[61]

Throughout the *Enneads* Plotinus employs the term *logos*, meaning "expressed principle," to indicate the image of the higher as it is found in the lower. He outlines the active presence of the *logos* (plural *logoi*) in nature as follows:

> In fact, the underlying and worked-upon matter comes to form bearing these [hot or cold], or becomes such when the expressed principle, though it itself does not have the property, works on it; for it is not necessary for fire to be added in order for matter to become fire, but rather an expressed principle [to be added], which is not an inconsiderable sign of both the fact that in living beings and in plants the expressed principles are the producers, and the fact that nature is an expressed principle, which makes another expressed principle, a product of it, giving something to the underlying subject, while it is itself static. (*Enneads* III.8.2)

In this way Plotinus refers to the rules or laws in the World-soul which are manifested in the "body" of nature.[62]

The process of emanation from the One via the Intellect and the Soul ends in matter (*hylē*). Since matter represents the final emana-

---

60. Moore, *Neoplatonism*; Moore, *Plotinus*.
61. Oosthuizen, J.S, *Van Plotinus tot Teilhard de Chardin. 'n Studie oor die metamorfose van die Westerse werklikheidsbeeld* [*From Plotinus to Teilhard de Chardin. A study on the metamorphosis of the Western world-view*] (Amsterdam: Rodopi N.V., 1974), 112, 114, 121.
62. Dillon and Gerson, *Philosophy*, 37.

tion from the One, it marks the point where the power of the One reaches a terminus.[63] Yet this recognition does not preclude Plotinus from holding an appreciative view of the sensible world. Because all things exist due to the power of the One, nothing prevents any of them from participating in the nature of the Good, according to the capacity of each. As he says in *Enneads* IV.8.6, "The most beautiful part of the sensible world, then, is a manifestation of the best among the intelligibles, of their power and of their goodness; and all things, both sensible and intelligible, are eternally connected, the intelligibles existing by themselves, the things that partake always receiving their existence from them, imitating the intelligible nature insofar as they are able." It was further argued by Plotinus that "ugliness is matter not conquered by form" (*Enneads* I.8.5). In other words, beauty is found in sensible things according to the measure of their participation in the intelligible Forms.

Of unknown authorship but attributed to Porphyry's student Iamblichus (c. 260–330) is a Neoplatonic work titled *The Theology of Arithmetic*, in which the author elaborates on the mystical, mathematical, and cosmological symbolism of the first ten numbers. It opens with the statement that the monad is the non-spatial source of number, and is called monad on account of its stability—the Greek *monas* deriving from *menein*, meaning to be stable. The monad is associated with God, who is seminally everything that exists, self-generated, and the cause of permanence in natures.[64] Iamblichus continues with the dyad, and associates it with matter as the source of differentiation in number and nature, respectively. The first odd number is the triad, which is the first number to indicate totality, since all natural process have a beginning, a middle, and an end. The tetrad is the first number to display the nature of solidity, the latter arising in the sequence point → line → plane → body. The tetrad thus provides the limit of corporeality and three-dimensionality. The pentad is the first number to encompass the

---

63. Oosthuizen, *Plotinus*, 131.

64. Iamblichus, *The Theology of Arithmetic. On the Mystical, Mathematical and Cosmological Symbolism of the First Ten Numbers*, trans. Robin Waterfield (Grand Rapids, MI: Phanes Press, 1988), 35, 37.

specific identity of all number, being the sum of the first even and uneven numbers (2 + 3). It is associated with all the natural phenomena of the universe, especially as it pertains to life.[65] In the latter regard one may consider, for instance, the pentagonal bodily form that underlies most terrestrial vertebrate life.

The works of the last major Neoplatonist, Proclus (412–485), comprise the most complete extant exposition of Platonism. His two major works of systematic philosophy and theology, namely *Elements of Theology* and *Platonic Theology*, continue the Neoplatonic synthesis of Plato, Aristotle, and the Stoics. The Latin and Arabic translations of these works became highly influential in medieval philosophy, both Christian and Islamic. In his *Elements of Theology*, Proclus conceives of universal nature as the result of a threefold "movement": the eternally abiding First Principle (*monē*), a procession (*proodos*) thereof through the Forms, and a return (*epistrophē*) through the Forms back to the First Principle. In other words, the cosmos comprises a ceaseless movement from Being into becoming and a return into Being.[66] For Proclus the dynamic nature of the cosmos is due to erotic love (*eros*) through which the multiplicity of things participate in their causes and ultimately in the One.[67] Here we encounter an echo of Empedocles' notion that the world of multiplicity arises through the agency of Love mixing and uniting unlike things.

Proclus' appreciative view of the physical world is evident from the following: "From this it clearly follows that even at the level of natural bodies unity and multiplicity coexist in such a manner that the one Nature contains the many natures as dependent on it, and, conversely, these are derived from one Nature—that of the whole" (*Elements of Theology*, Proposition 21). Ultimately, Nature is rooted in the One, as Proclus asserts: "That, then, is the One the principle of all generation both for the manifold powers of nature, and for

---

65. Ibid., 40, 42–43, 49–51, 55, 58, 63, 65–66.
66. Sheldon-Williams, *Platonist*, 431; Moore, *Neoplatonism*.
67. Goosen, Danie, *Die Nihilisme. Notas oor ons tyd* [Nihilism. Notes on our time] (Pretoria: Praag, 2007), 42.

Aspects of Hellenic Cosmology

particular natures, and for all those things under the sway of nature" (*Commentary on Parmenides*, VI 1046).[68]

Since matter is moved by soul, we should briefly consider the psychology of Proclus. In his *Elements of Theology* he states that every soul is an incorporeal substance and separable from the body (Proposition 186), is indestructible and imperishable (Proposition 187), contains all the Forms that Intellect contains primarily (Proposition 194), and is all things, both sensible entities in the mode of an exemplar and intelligibles in the mode of an image (Proposition 195). Since Soul proceeds from Intellect, the individual soul possesses the reflections (*emphaseis*) of the Forms, albeit in a secondary mode. Moreover, because soul participates in Intellect, it is able to impart the reason-principles (*logoi*) of material things, thereby causing bodies to exist. Finally, against Plotinus' notion that a part of the individual soul remains above the sensible realm, Proclus argues in favor of full indwelling: "Every particular soul, when it descends into the realm of generation, descends completely; it is not a case that there is a part of it that remains above and a part that descends" (*Elements of Theology*, Proposition 211).[69] In this way physical bodies receive both their being and intelligibility from the *logoi* provided by their souls.

Regarding epistemology, Proclus follows Empedocles and Plato in asserting that at every level of reality, like is known by like: the sensible realm is known by sensation (*aisthēsis*), the heavenly realm by opinion or belief (*doxa*), the realm of Soul by discursive reasoning (*dianoia*), and the intelligible realm by intellection (*noēsis*). When the soul contemplates the universe, it sees only the images of true beings, but when it turns into itself the soul perceives its own reason-principles (*logoi*), the latter being the projections within the soul of the eternal Forms. In the words of Proclus, "For all things are within us in a manner proper to soul, and by reason of that we have the natural capacity to know everything through awaking the powers within us and the images of all beings" (*Platonic Theology*

---

68. Dillon and Gerson, *Philosophy*, 271, 326.
69. Ibid., 278–80.

I.3).[70] According to the Platonic hierarchy of knowledge, opinion is the lowest power of the rational soul, limited to knowledge of the universal in particulars. That is to say, opinion can only know that a thing is (*oti*), but not why it is (*dioti*).[71] Intellection, in contrast to discursive reasoning that strives toward rational, "scientific" knowledge, is non-discursive knowledge which implies the soul's affinity with the Forms (*ta noēta*).[72] Given its firm ontological basis, we find Neoplatonic epistemology superior to most modern epistemology, which lacks a sound ontological foundation.

---

70. Ibid., 291.
71. Taylor, Thomas, *Introduction to the Philosophy & Writings of Plato* (Seaside, OR: Watchmaker Publishing, 2010), 105.
72. Uzdavinys, Algis, *Orpheus and the Roots of Platonism* (London: The Matheson Trust, 2011), 76.

3

Hellenic Philosophy of Life, or Bio-Philosophy

ALTHOUGH ARISTOTLE is the undisputed colossus among the Hellenic philosophers regarding biological thought, some of his predecessors also made notable contributions. It has been suggested that Anaximander's (c. 610–546 BC) notion of rhythmic balance among the primary elements of water, air, fire and earth anticipates the twentieth-century theory of dynamic equilibrium in ecosystems. According to this theory, creation, destruction, regeneration and decomposition occur continuously and succeed one another periodically. New organisms are born when old ones die, composed out of the dissolved materials of dead organisms. Anaximander also recognized that the generation and destruction of things occur "in accordance with the ordering of time," which fits with the observed rhythmic cycles of biological systems, with populations alternately increasing and decreasing in number.[1]

Anaximander explained the origin of living beings in terms of new, more complex things arising out of simpler existing things. According to the Roman grammarian Censorinus (third century AD), Anaximander held that life originated from heated water and earth, which gave rise to fish-like beings. Inside these aquatic creatures humans grew as embryos until they reached puberty, whereupon the containers burst open and men and women came forth. This notion of Anaximander could be viewed as an incipient evolutionary understanding of the organic world, in terms of which all life began in the sea, and later terrestrial animals (including

---

1. Theodossiou et al., *Theories*, 94.

From Logos to Bios

humans) developed from aquatic animals.² If we add that this development takes places according to inherent possibilities, i.e., in a directed manner, we might say that Anaximander depicted in mytho-poetical language the traditional understanding of evolution, namely the unfolding of that which is in-folded.

The cosmogony of Pythagoras and his followers is also relevant to the notion of evolution according to natural law. In his *Pythagorean Notebooks*, the Hellenic scholar Alexander Polyhistor (c. 105–c. 35 BC) summarizes the sequence of generation from numbers to the cosmos as follows: from the unit and the indefinite dyad arise numbers, which give rise to points, which give rise to lines, which give rise to plane figures, which give rise to solid figures, which give rise to sensible bodies. Sensible bodies consist of the four elements fire, water, earth and air, and these elements "interchange and turn into one another completely, and combine to produce a universe (*kosmos*) animate, intelligent, spherical, with the earth at its center, the earth itself being spherical, and inhabited round about."³ Incidentally, this quote is a reminder that the Hellenic thinkers were familiar with the notion of a spherical Earth, as also were Plato and Aristotle.

This geometrical cosmogony of the Pythagoreans could equally well be stated in terms of an evolutionary coming-to-be. James Cutsinger writes that the point harbors an (innate) tendency to "spill over the edges of its invisible and dimensionless essence into the dimensions of the things that are seen," thereby producing a line. But since the latter only actualizes the possibilities of the point in one dimension, length, the point seeks a further dimension of breadth, and thus the plane is evolved. However, since the two-dimensional plane is still invisible and incorporeal, the point evolves or "explodes" into a third dimension of depth, whereby the solid is produced. In this way the terrestrial world of material objects comes to be, which is the domain of the evolution of species.⁴

The first Athenian philosopher, Anaxagoras (c. 500–428 BC),

2. McKirahan, *Philosophy*, 42; Swift, *Evolution*, 58.
3. McKirahan, *Philosophy*, 101.
4. Cutsinger, *Earth*, 13–14, 23.

postulated an infinite number of elements as first principles, rather than holding any one element as the single first principle or *archē*, as we are told by Aristotle (*Metaphysics* I.3, 984a). Each of these elements, such as water, gold, or blood, is identical in all its parts, and each is infinitely divisible into minute particles called *chremata* ("things") or *spermata* ("seeds"). Some quality of each *sperma* is to be found in all other *spermata*, but its own properties remain dominant. This is reminiscent of the claim in molecular biology that all living beings on Earth are genetically related, since all living organisms develop and reproduce according to the instructions in their deoxyribonucleic acid molecules, or DNA. Moreover, Anaxagoras' idea that all things are composed of identical parts has been interpreted as implying that change involves the transmission of something that is already present. This insight would exercise a lasting influence in the metaphysics of causality, notably in biology.[5] It is also relevant to the theory of evolution according to natural law, which teaches that life-forms unfold as determined by their inherent possibilities.

We next move to Empedocles (c. 495–435 BC), who maintained that the generation of animals takes place in stages, which occur within the larger context of the cosmic interaction between Love and Strife.[6] The process begins in a period of increasing Love, so that the elements are united into body parts, but the dominance of Strife prevents these parts from joining together to form animals: "By her [Love] many neckless faces sprouted, and arms were wandering naked, bereft of shoulders, and eyes were roaming alone, in need of foreheads" (Fragment 57), and, "In this situation, the members were still single-limbed as the result of the separation caused by Strife, and they wandered about aiming at mixture with one another" (Fragment 58). In the second stage a greater inter-mixture of body parts becomes possible as Love's influence increases further: "But when divinity was mixed to a greater extent with divinity, and these things began to fall together, however they chanced to meet, and many others besides them arose continuously" (Fragment 59).

---

5. Dreyer, *Wysbegeerte*, 53; Blackburn, *Philosophy*, 15.
6. McKirahan, *Philosophy*, 278.

However, due to the presence of Strife this was still an apparently random process, resulting in monstrous human-animal combinations familiar in Hellenic mythology: "Many came into being with faces and chests on both sides, man-faced ox-progeny, and some to the contrary rose up as ox-headed things with the form of men, compounded partly from men and partly from women, fitted with shadowy parts" (Fragment 61). Nevertheless, in this latter stage limbs began to form viable combinations, representing the beginning of the natural history of living animal species. Viability in this regard pertains not only to individual survival, but even more so to reproduction, which enables continuance of species without resorting to chance combinations. This perpetuation of species represents a further advance of Love.[7] By postulating that viable combinations of limbs and organs survive while unsuitable combinations perish, Empedocles could be said to have anticipated the modern evolutionary notion of natural selection and survival of the fittest.[8]

Plato

In Plato's dialogue *Timaeus*, we read that the Demiurge or divine Craftsman made four kinds of living beings corresponding to the four primary elements of fire, air, water, and earth (39e–40a). The four kinds of beings are: (i) the heavenly gods, in which the element of fire dominates; (ii) flying creatures (air); (iii) aquatic creatures (water); and (iv) terrestrial creatures (earth). It should be noted that the Demiurge himself makes only the heavenly gods, while these gods in turn make the remaining three classes of living beings. Plato's delegation of the rest of the creative work to the celestial gods may reflect a notion that the heavenly bodies, especially the Sun, actively generates life on Earth.[9] Thus in the *Republic* the Sun is named as the cause of coming to be, growth, and nourishment of things in the visible world, without itself coming to be (Book VI, 509b). Plato's reference in the *Timaeus* (40b) to "living creatures everlasting and divine" includes the fixed stars and the planets, of

---

7. Ibid., 245–46, 279.
8. Swift, *Evolution*, 58.
9. Cornford, *Timaeus*, 118, 141.

Hellenic Philosophy of Life, or Bio-Philosophy

which the Earth (*Gaia*) is the one with greatest seniority (40c). This statement has momentous ecological implications, since according to this view the Earth is then not a lifeless object to be plundered and desecrated by humankind as it has often been regarded in the modern industrial world.

The instruction of the Demiurge to the gods, "Weave what is mortal to what is immortal, fashion and beget living things" (*Timaeus* 41d), is followed by the creation of human beings. However, since anthropology falls outside the scope of this book, we will not consider this aspect of Plato's cosmology. Of more relevance to us is that following his fairly detailed account of the creation of the human being, Plato describes its relation with the plant kingdom:

> They [i.e., the gods] made another mixture and caused another nature to grow, one congenial to our human nature though endowed with other features and other sensations, so as to be a different living thing. These are now cultivated trees, plants and seeds, taught by the art of agriculture to be domesticated for our use. But at first the only kinds there were wild ones, older than our cultivated kinds. We may call these plants "living things" (*zōa*) on the ground that anything that partakes of life has an incontestable right to be called a "living thing." (*Timaeus* 77a–b)

This reasoning implies that not only animals but plants also are in possession of soul, since it would be impossible to be a living thing without a soul. In the final sentence of the dialogue's penultimate paragraph, Plato says: "Thus, both then and now, living creatures keep passing into one another in all these ways, as they undergo transformation by the loss or by the gain of reason and unreason" (92b–c).[10] When this statement is viewed together with the earlier passage on the creation of plants, Plato seems to conclude that there is no firm division between the three main categories of living creatures, namely humans, animals, and plants—another insight that has been confirmed by modern biology, namely of a shared genetic inheritance among all living beings.

---

10. Translation in Theodossiou et al., *Theories*, 96.

Aristotle

Aristotle is widely regarded as the father of biological studies in the Western world. This recognition dates back to the classical Hellenic era, with its saying that "Aristotle was nature's scribe, his pen dipped in mind (*nous*)."[11] His biological writings are infused with metaphysical insights, in contrast to modern biology, with its insistence that metaphysical concepts are outside its scope. Aristotle viewed biology and psychology not as separate sciences, but rather as a single science of life. He wrote several works that dealt with life science, including *The History of Animals*, recording facts about hundreds of animal species, *On the Parts of Animals*, which dealt with the material qualities of animals, *On the Soul*, which discussed their essential form, and *On the Generation of Animals*, concerning procreation.[12] The importance of the study of animals is eloquently presented by Aristotle in his treatise *On the Parts of Animals*, where he writes:

> For if some [i.e., animal species] have no graces to charm the sense, yet even these, by disclosing to intellectual perception the artistic spirit that designed them, give immense pleasure to all who can trace links of causation, and are inclined to philosophy.... Every realm of nature is marvelous ... so we should venture on the study of every kind of animal without distaste; for each and all will reveal to us something natural and something beautiful. (I.645a)

With this statement, Aristotle also makes clear that his approach to biology is firmly grounded in formal and final causality.

In his *Physics*, Aristotle, makes a fundamental distinction between the natural and the artificial, resulting from the kind of movement (*kinēsis*) a being exhibits. Natural things have their principle of movement within, while for artificial things, the source of movement is external to them (*Physics* II.192b). Building on this differentiation, the biochemist Michael Denton and the geneticists Craig Marshall and Michael Legge, reason that the realm of the natural entails a finite number of forms, but an infinite number of paths toward actualization. In contrast, the realm of the artificial entails an

---

11. Quoted in O'Rourke, *Metaphysics*, 3.
12. Ross, *Aristotle*, 117.

infinity of forms, but each of these forms is assembled by only a few or even only one process, as is the case with machines and watches.[13] This suggests that, as a natural phenomenon, the evolutionary process is based on a finite number of intelligible forms (*eidai*), which are manifested in an almost infinite variety of morphological forms (*morphai*). Already in the nineteenth century it had been suggested by the English naturalist Alfred Wallace that the virtually infinite variety observed in nature can be traced back to the almost infinite complexity of the cells that constitute living beings, of the protoplasm which is the substance of the cells, of the elements that constitute the protoplasm, of the molecules of those elements, and finally of the atoms whose combinations form those molecules.[14] From this, it would follow that the vast morphological variety observed in the living kingdoms is at least partially a result of biochemical complexity.

Form

In contrast to the materialist bias that has dominated the scientific world since the seventeenth century, Aristotle insisted that the nature of things is more properly found in their form (*eidos* or *morphē*) than in their matter (*hylē*): "For the formal nature is of greater importance than the material nature" (*Parts of Animals* I.640b). If this is so, then Democritus was wrong to claim regarding human beings that the essence of animals and their parts is to be found in their configuration or color, for, as Aristotle argues, a dead body has the same configuration as a living one, but he is decidedly not a man (*Parts of Animals* I.640b). Aristotle's understanding of the body-form relation would later be accepted by Thomas Aquinas, who, in his treatise *On Being and Essence*, writes that "For the animating soul isn't a different form from the one enabling the thing to occupy three dimensions, and when we said body is anything of a

---

13. Denton, Michael J. et al., "The Protein Folds as Platonic Forms. New Support for the pre-Darwinian Conception of Evolution by Natural Law," in *Journal of Theoretical Biology* 219 (2002): 333.

14. Flannery, Michael A., *Alfred Russell Wallace's Theory of Intelligent Evolution* (Riesel, TX: Erasmus Press, 2011), 196–97.

form such that it can occupy three dimensions we were meaning whatever form that may be.... And in this way the animal form was implicit in the form of body."[15]

Aristotle follows Plato in holding that form expresses all that is intelligible in nature, while matter, in contrast, is only an object of knowledge by analogy, as bronze is to a statue or wood to a bed (*Physics* I.191a).[16] Now Aquinas holds that the combination of matter and form in the establishment of substance applies to the levels of the individual, the species, and the genus, so that the matter and form of the same individual (say Cicero) are the same individually; the matter and form of different individuals of the same species (say Socrates and Plato) are the same in species but differ individually; and the bodies and souls of donkeys and horses, for example, differ in species but are the same generically (*On the Principles of Nature*). In other words, as he writes elsewhere, form demarcates species within a genus, whereas matter demarcates individuals within a species (*On Being and Essence*).[17] This understanding of form and matter and how each relates to species, contradicts those who deny the reality of species, something we will discuss in more detail in a later chapter. For now, let us just observe that in his *History of Animals*, Aristotle had already introduced the concept of species as natural kinds that reproduce true to their nature,[18] which anticipates the modern biological definition of a species as comprising all those organisms capable of producing fertile offspring.

Aristotle's insistence on the priority of form over matter in determining the nature of a thing is related to his metaphysics of potency and actuality. As he reasons in the *Physics*, "The form indeed is 'nature' rather than the matter; for a thing is more properly said to be what it is when it has attained to fulfilment than when it exists potentially" (II.193b), or, to rephrase, a thing has its nature more fully when it exists actually (i.e., has attained its form) than when it

---

15. McDermott, *Aquinas*, 95–96.
16. Gerson, *Aristotle*, 240.
17. McDermott, *Aquinas*, 79–80, 98.
18. Swift, *Evolution*, 59.

exists potentially (i.e., only the matter exists).[19] Later in the *Physics*, Aristotle emphasizes the continuity that is present in the transition from potency to actuality: "Since everything that changes changes from something to something, that which has changed must at the moment when it has first changed be in that to which it has changed" (VI.235b). Elsewhere he argues along similar lines: "Reflection confirms the observed fact; the actuality of any given thing can only be realized in what is already potentially that thing, i.e., in a matter of its own appropriate to it" (*On the Soul* II.414a). In the case of living beings, therefore, the development of an organism can only be actualized according to its inner potentiality. This view is clearly at odds with the Darwinian hypothesis that new species are formed due to material and mechanical factors alone.

In a remarkable anticipation of the modernist claim that the evolution of life from simpler to more complex organic forms rebuts the traditional notion of formal causality, Aristotle states that the order of actual development of things and the order of logical existence are always the inverse of each other (*Parts of Animals* II.646a). In other words, that which is posterior in the order of development is antecedent in the order of nature. Aristotle illustrates his argument by pointing to the fact that a house does not exist for the sake of bricks and stones, but these materials exist for the sake of the house, and therefore "in order of time, the material and the generative process must necessarily be anterior to the being that is generated; but in logical order the definitive character and form of each being precedes the material" (*Parts of Animals* II.646a–b). Aquinas makes a similar distinction between the temporal process of generation and the completeness of being, saying that adults are prior to children in completeness of being, while children are prior to adults in the temporal process of generation (*On the Principles of Nature*).[20] This is sufficient to show that the evolution of life from simpler to more complex forms does not refute the priority of form over matter.

19. Ross, *Aristotle*, 71.
20. McDermott, *Aquinas*, 75.

Reproduction and Growth

For Aristotle reproduction is the most fundamental phenomenon of life, even more so than movement and sensation, since it is the only bio-phenomenon that can occur without the others, as is the case among plants.[21] As he states in the *History of Animals*, "Thus of plants that spring from seed the one function seems to be the reproduction of their own particular species, and the sphere of action with certain animals is similarly limited. The faculty of reproduction, then, is common to all alike" (VIII.588b). Aristotle provides an ontological and psychological grounding for the striving of all living beings to reproduce their type faithfully:

> And since soul is better than body and the ensouled is better than the soulless owing to its soul and being (*to einai*) is better than not being (*me einai*) and living (*to zen*) better than not living (*me zen*), for these reasons reproduction (*genesis*) of living things exists.... Since it is impossible for it [i.e., a living thing] to be eternal as an individual..., it is possible for it to be eternal in species. This is the reason why there exists eternally the class of human beings, animals, and plants. (*Generation of Animals* I.731b–732a)[22]

That is to say, because individuals are unable to live eternally, but rather succumb to death and decay, it is the class (*genos*) and species (*eidos*) of organisms that are self-perpetuating by means of reproduction.[23]

Aristotle further applies the roles of form and matter to sexual differentiation: whereas the female is the material principle in sexual reproduction, the male is the formal principle. That is to say, the male parent impresses a certain form on the matter supplied by the female parent. The male semen thus acts as formal cause of the offspring, while the female menstrual discharge acts as material cause (*Generation of Animals* I.729a–730b).[24] The biophysicist Max Delbrück suggests that with his notion of the male semen providing the

---

21. Ross, *Aristotle*, 122.
22. Translation in Gerson, *Aristotle*, 118.
23. O'Rourke, *Metaphysics*, 46.
24. Ross, *Aristotle*, 123–24.

Hellenic Philosophy of Life, or Bio-Philosophy

formal principle in the generation of animals, Aristotle discovered the principle implied in DNA. Accordingly, "The form [sic] principle is the information which is stored in the semen. After fertilization it is read out in a preprogrammed way; the readout alters the matter upon which it acts, but it does not alter the stored information, which is not, properly speaking, part of the finished product."[25]

In his *Generation of Animals*, Aristotle rejects the Hippocratic theory of pangenesis, according to which the embryo contains all its parts fully preformed in miniature. In its place, he offers a theory more in line with his metaphysics. Fran O'Rourke summarizes it as follows: "The parts of the animal are formed successively, with the gradual actualization of what is initially present in potency, under the agency of what is actual." Aristotle thus provides the conceptual foundation for the empirically verified theory of epigenesis, which holds that "embryonic development is a chain of new constructions, each perfecting the preceding, with the final differentiation of the living individual emerging at the end."[26] To a certain extent, then, the science of embryology thus appears to have preserved teleology.[27]

Aristotle recognized that heredity is related to procreation by the sexual act, writing: "In animals where generation goes by heredity, wherever there is duality of sex generation is due to copulation" (*History of Animals* V.539a). He adds that "man is generated from man; and thus it is the possession of certain characters by the parent that determines the development of like characters in the child" (*Parts of Animals* I.640a), whether the "characters" be from the father or from the mother. If the impulses imparted by the parents are confused, then the offspring will be unlike the parents, but they will yet preserve the character of the species.[28] Remarkably, Aristotle's thoughts on heredity anticipated the modern sciences of molecular biology and genetics. According to the philologist Wolf-

---

25. Quoted in O'Rourke, *Metaphysics*, 12.
26. O'Rourke, *Metaphysics*, 9–10.
27. Thompson, *Growth*, 4.
28. Ross, *Aristotle*, 127.

gang Kullmann, "Aristotle's genetics... has an extraordinary similarity with the modern theories in molecular biology of DNA and the genetic code. Aristotle's position... is more balanced than the picture of embryology and genetics in the first half of the twentieth century."[29]

It is significant that Aristotle does not view the growth of organisms as undirected—to the contrary, every growth has a purpose (*telos*) towards which it proceeds.[30] Thus, in the *Parts of Animals*, it is stated, "Again, whenever there is plainly some final end, to which a motion tends should nothing stand in the way, we always say that such final end is the aim or purpose of the motion" (I.641b). Elsewhere Aristotle says that the development of living beings is preceded by their essence: "The ordered and definite works of nature do not possess their character because they developed in a certain way. Rather they develop in a certain way because they *are* that kind of thing, for development depends on the essence and occurs for its sake. Essence does not depend on development" (*Generation of Animals* 778b).[31] Stated in terms of causality, the growth and development of organic entities is attributable to both formal and final causes.

Soul

Soul is defined by Aristotle as "the first grade of actuality of a natural body having life potentially in it. The body so described is a body which is organized" (*On the Soul* II.412a). A useful alternative translation of this definition has been provided by Lloyd Gerson, who writes that the soul is "the primary actuality (*entelecheia*) of a natural body with organs (*sōmatos physikou organikou*)."[32] According to Aristotle, the relation of soul to body is that of actuality to potentiality, of form to matter. And since substance consists of both form and matter, a particular soul is inseparable from its body; in other words, the soul is the actuality of a certain kind of body (*On the*

---

29. Quoted in O'Rourke, *Metaphysics*, 11–12.
30. O'Rourke, *Metaphysics*, 36.
31. Translation in O'Rourke, *Metaphysics*, 41.
32. Gerson, *Aristotle*, 133.

Soul II.413a). Further on in the same treatise, Aristotle says that the soul is the cause of the body in three senses: (a) as the source or origin of movement, whether it be locomotion or alteration; (b) as the end or final cause; and (c) as the essence of the whole living body (*On the Soul* II.415b). In other words, the soul is the form of a living being, guiding the latter towards its purpose, and each living thing thus has a natural *telos*.

Aristotle lists five powers of the soul: nutritive, appetitive, sensory, locomotive, and cognitive. Plants have nutritive power, while animals have that as well as appetitive, sensory, and locomotive powers. Humans have all these powers, plus the cognitive, i.e., mind (*On the Soul* II.414a–b). Since the nutritive faculty is the most widely distributed power of soul it exists in all living beings from birth until death. It manifests itself in the use of food and in reproduction (*On the Soul* II.415a, III.434a). For all living things whose mode of generation is not spontaneous, "the most natural act is the production of another like itself, an animal producing an animal, a plant a plant, in order that, as far as its nature allows, it may partake in the eternal and divine" (*On the Soul* I.415a–b). This notion of participation in the divine implies that the ultimate aim of nutrition and reproduction is the preservation not of individual life, but of the species.[33] Plato likewise asserts that the mortal nature of both humans and animals strives towards immortality by means of reproduction, which is the only means possible for it (*Symposium* 207c–d).[34]

For Aristotle, nutrition and reproduction are due to the same power of soul. Accordingly, what is fed is the "besouled body" (*empsychon soma*), and thus food is essentially related to what has soul in it. Stated the other way round, if the nutritive soul is deprived of food, it ceases to be (*On the Soul* II.416a–b). Since nutrition and reproduction are psychically linked, the life of animals is concentrated on these two activities. Aristotle adds that "whatsoever is in conformity with nature is pleasant, and all animals pursue pleasure in keeping with their nature" (*History of Animals* VIII.589a). Inter-

---

33. Ross, *Aristotle*, 141.
34. Gerson, *Aristotle*, 118.

estingly, the mutual friendship or enmity between various animal species is attributed to the food they feed on and the life they lead (*History of Animals* IX.610a).

Next in Aristotle's schema, we have the sensitive power of the soul, which is common to all animals but not to plants. Each of the five senses has the power of receiving into itself the sensible forms of things without the matter. Of these powers, touch is indispensable to all animals, since its loss would bring about the death of the organism. The remaining senses serve well-being rather than mere survival, as Plato had also argued, since sight enables an animal to see through air or water, taste enables it to distinguish painful or pleasant qualities in its nutrition, and hearing enables communication with its fellows (*On the Soul* II.424a, III.434b, 435b). Aristotle adds that in the case of humans, sight and hearing also pertains to the life of thought, in that the hearing of speech is the main instrument of teaching and learning, while sight reveals differences in color, number, size, shape and movement (*Metaphysics* I.980a–b).

The highest level of soul is the rational, which is found among human beings alone, and Aristotle with admirable modesty admits that the questions of when, how, and whence regarding the acquisition of mind by humans is a most difficult one (*Generation of Animals* 736b).[35] For Aristotle the rational function differs from the other powers of soul, since "mind must be related to what is thinkable, as sense is to what is sensible." The rational soul is therefore potentially capable of contemplating the forms of existing things (*On the Soul* III.429a). Against the modern reductionist fantasy that mind is an epiphenomenon of the brain, Aristotle argued that mind cannot reasonably be viewed as blended with the body, since it cannot acquire a quality such as hot or cold, or an organ of sense-perception. He also writes that "since it is the soul by or with which primarily we live, perceive, and think:—it follows that the soul must be a ratio or formulable essence, not a matter or subject" (*On the Soul* II.414a). In the Aristotelian conception, reason has no connection with matter and enters it from the outside, being divine in

---

35. O'Rourke, *Metaphysics*, 31.

nature (*Generation of Animals* 736b). Aristotle further draws a distinction between the sensitive soul and the rational soul, which leads to the distinction between sense perception (*aisthēsis*) and contemplative thought (*noēsis*). This distinction would be taken over by Plotinus and become normative in the Neoplatonic epistemology.[36]

Aristotle's Scale of Nature

Probably spurred by his wide-ranging biological interests, Aristotle became the first Western thinker to attempt a classification of animal species. He recognized three grades of likeness within the animal kingdom: identity likeness among individuals within a species, likeness among species of the same genus, and likeness among "greater genera"—for instance, the homology between arm, foreleg, wing, and fin among various vertebrate classes.[37] The guiding principle for Aristotle's classification of animals is stated in the *Parts of Animals*: "Groups that only differ in degree, and in the more or less of an identical element that they possess, are aggregated under a single class; groups whose attributes are not identical but analogous are separated" (I.644a). Whereas a species (e.g., human) is characterized by all its individual members possessing common attributes, a larger group (e.g., birds or fishes) is determined by a similarity in the shape of particular organs or of the whole body (*Parts of Animals* I.644a–b). Aristotle thus anticipated the taxonomical project of Carl Linnaeus by more than two millennia.

Animals were arranged by Aristotle in a hierarchy according to the degree of development reached by the offspring at their time of birth (*Generation of Animals* II.732a–733b). This degree of development was ascribed to the degree of vital warmth possessed by the parent animals. In this scheme, the highest types of animals are the viviparous ones, among which the offspring are smaller versions of the parents. The next highest are egg-laying: first those types producing a "perfect" egg, i.e., the egg does not grow in size after being laid; then the types producing an "imperfect" egg, which has to

---

36. Oosthuizen, *Plotinus*, 101.
37. Ross, *Aristotle*, 118–19.

grow. The whole scale of nature then appears as follows, in descending order: humans, land mammals, sea mammals, birds, reptiles and amphibians, fishes, cephalopods, crustaceans, insects, mollusks other than cephalopods, and finally zoophytes.[38]

As a keen observer of nature, Aristotle also recognized the existence of intermediate life-forms. In the *History of Animals* he depicts the continuity between plants and animals, and between the inanimate and the animate:

> Nature proceeds little by little from things lifeless to animal life in such a way that it is impossible to determine the exact line of demarcation, nor on which side thereof an intermediate form should lie. Thus, next after lifeless things in the upward scale comes the plant, and of plants one will differ from another as to its amount of apparent vitality; and, in a word, the whole genus of plants, whilst it is devoid of life as compared with an animal, is endowed with life as compared with other corporeal entities. Indeed, as we just remarked, there is observed in plants a continuous scale of ascent towards the animal. So, in the sea, there are certain objects concerning which one would be at a loss to determine whether they be animal or vegetable. (VIII.588b)

This same idea is affirmed again in the *Parts of Animals*: "Nature passes in a continuous gradation from lifeless things to animals, and on the way there are living things which are not actually animals, with the result that one class is so close to the next that the difference seems infinitesimal" (681a).[39]

Considered in conjunction with his doctrine on the serial order of souls (i.e., nutritive, sensitive, and rational), Aristotle's ascending scale of nature has been viewed by some modern thinkers, for example the British scientist and historian, Joseph Needham, as anticipating the notion of evolution, once a temporal dimension is added. However, as the philosopher Fran O'Rourke correctly notes, the Darwinian conception of evolution due to strictly material and external factors would undoubtedly have been rejected by Aristotle.

---

38. Ibid., 120–22.
39. Translation in O'Rourke, *Metaphysics*, 39.

Hellenic Philosophy of Life, or Bio-Philosophy

In addition, although Aristotle recognized the existence of intermediate forms, his hierarchy of being made no allowance for evolutionary progression through transformism.[40]

Ultimately, Aristotle saw the world of living beings as integrally connected. In the *Metaphysics* he declares, "And all things are ordered together somehow, but not all alike—both fishes and fowl and plants; and the world is not such that one thing has nothing to do with another, but they are all connected. For all are ordered together to one end . . . that all must at least come to be dissolved into their elements, and their other functions similarly in which all share for the good of the whole" (XII.1075a). The interconnectedness of all living things is grounded in teleology, being ordered toward a specific end.

Aristotle's Teleology

By way of clarification, we should note that Aristotle does not use the term "final cause," as such, but rather the "end" (*telos*), the "in view of which" (*to hou heneka*), and the "why" (*dia ti*) of things.[41] He states repeatedly that all things have a reason for being: "For Nature, like mind, always does whatever it does for the sake of something, which something is its end" (*On the Soul*, II.415b); "Nature never makes anything without a purpose and never leaves out what is necessary, except in the case of mutilated or imperfect growths" (III.432b); and "For all things that exist by Nature are means to an end, or will be concomitants of means to an end" (III.434a), and, finally, "Everything that Nature makes is means to an end" (*Parts of Animals* I.641b). In other words, everything in the natural world is ordered to a certain purpose, although exceptions such as physical deformities do occur.

Aristotle's teleology enables him to indemnify the Prime Mover from the imperfections in nature. Thus imperfections in the structure of animals are ascribed to defective material, not a defective maker: it can happen that matter is sometimes not suitable for the purpose in hand. Imperfections in individual organisms are due to

---

40. O'Rourke, *Metaphysics*, 39, 41; Swift, *Evolution*, 60.
41. Gilson, *Aristotle*, 5.

the inherent variability of matter, since the latter is formed of an endless variety of combinations of the four elements.[42] Nevertheless, Aristotle emphasizes that final causality is paramount: "Both causes must be stated by the physicist, but especially the end; for that is the cause of the matter, not vice versa" (*Physics* II.200a).

Aristotle's teleology is integrally linked to his metaphysics of causality. This enabled him, for instance, to distinguish between characteristics on the species level and on the individual level. Qualities that characterize the whole of a species are to be explained by final and formal causes, while variable characteristics are to be explained by material or efficient causes. For example, in the *Generation of Animals*, Aristotle suggests that the formation of a person's eye serves a certain purpose in accordance with the reason (*logos*) of the individual, while the color of the eye is incidental and must "of necessity" (*ex anangkes*) be ascribed to its matter and moving cause (778a–b). That is to say, spontaneous variations among individual organisms should be explained by material and efficient causes rather than final causes.[43]

The implications of Aristotle's scheme of causality for molecular biology has been deftly outlined by David Swift in his critique of evolution. Swift writes that the material cause of biological macromolecules is the constituent atoms, the formal cause is the proteins or genes, and the final cause is their role in biological systems, such as to code for a protein or RNA, or to build a tissue. However, Swift continues, an efficient cause for biological macromolecules is lacking, since their immense complexity (which will be discussed later in this book) cannot arise through a chance event or a progressive selective process, as is postulated in the Neo-Darwinian model. Swift concludes, "Biological macromolecules, in themselves, present a case for design for which we do not have a natural or scientific explanation; they point clearly to there having been a purposeful designer."[44]

---

42. Ross, *Aristotle*, 130.
43. O'Rourke, *Metaphysics*, 42; Ross, *Aristotle*, 127.
44. Swift, *Evolution*, 398–99.

Hellenic Philosophy of Life, or Bio-Philosophy

Final causality also comes into play in Aristotle's differentiation between homogeneity and heterogeneity in the body parts of animals. Aristotle outlines three degrees of composition of body parts: (i) primary substances are composed out of the elementary forces of earth, air, water, and fire; (ii) homogeneous parts, such as bone and flesh, are composed out of the primary substances; (iii) heterogeneous parts, such as face and hands, are composed out of homogeneous parts. Homogeneous parts are therefore ordered to heterogeneous parts as to a final cause (*Parts of Animals* II.646a–b). All the living operations of animals and plants require differentiation of parts that are capable of interacting. Heterogeneous parts require organization, which is why living bodies are called organisms or living matter is organic. For Aristotle, mechanical explanations in terms of material and efficient causality is sufficient to explain homogeneous parts, but organisms with heterogeneous structures require a more complex explanation, one that involves final causality.[45]

For Aristotle, the teleology of nature is considerably more perfect than the teleology of art. Whereas the artist flounders, nature, although limited by matter, generally attains her end without hesitation. Accordingly, as Gilson observes, "Aristotle conceives the artist as a particular case of nature," which is why "art imitates nature, rather than nature imitating art."[46] In the Aristotelian conception, there is more design (*to hou heneka*), more good (*to eu*), and more beauty (*to kalon*) in the works of nature than in those of art. As he says in *Generation of Animals*, "In all this Nature acts like an intelligent workman" (I.731a). While art has its cause in the human intelligence, teleology in nature is a mystery to us, but for Aristotle its existence cannot be denied.[47] That the activity of nature is directed towards order and beauty is repeatedly affirmed by Aristotle: "But that which is produced or directed by nature can never be anything disorderly: for nature is everywhere the cause of order" (*Physics* VIII.252a); "Nature always strives after the better" (*On Generation*

---

45. Gilson, *Aristotle*, 4–7.
46. Ibid., xvi.
47. Ibid., 12–13.

and Corruption II.336b); and, "Nature ever seeks amend" (*Generation of Animals* I.715b).

It has been noted that for Aristotle the *telos* of living beings does not arise from the physical order of nature (*physis*). Rather, *telos* signifies the limit of living beings that from the beginning presides over the ordering of their parts, resulting in the establishment of beauty and order. Aristotle holds that the beauty of a living organism has nothing to do with utility, since beauty is an end in itself and not a means to something else.[48] This Aristotelian notion of beauty as an end in itself has been evoked by the British traditionalist Lord Northbourne in a delightful essay titled "Flowers," in which he notes that in the modern scientific view a flower is merely a mechanism for the transfer of pollen to another flower of the same species. The form, color and fragrance of flowers are claimed to have evolved to attract insects, thus compensating for the immobility of plants. However, as Northbourne notes, there exist many species of flowering plants in which the size or brilliance of the flowers bears no relation to their attractiveness to insects, so that the beauty of flowers cannot be the result of utilitarian causes. The phenomenon of floral beauty rather displays the underlying harmony of the created order, obscured by the prevalence of struggle. Since struggle is not the fundamental force shaping the natural world, says Northbourne, "still less did it produce the beauty of flowers, as is postulated in [Darwinian] evolutionist theories."[49] We submit that the beauty of flowers is neither accidental nor limited to utilitarian purposes; rather, the manifold occurrence of beauty in the organic realms is attributable primarily to final causality and not to mechanism.

The biological implications of Aristotle's insistence on both material and formal causality in the plant and animal kingdoms have more recently been lucidly treated by the Catholic philosopher Michael Tkacz:

> The form of organisms and the natural agencies that produce those forms can only be understood in relation to the material

---

48. Ibid., 19, 24.
49. Lord Northbourne, *Looking Back on Progress*, 90, 93, 98.

constitution of organisms. At the same time, the form of an organism cannot be reduced to its matter and the final optimal state of the organism cannot be reduced to the natural processes that bring it about. Organic morphology remains irreducibly formal while existing only in matter. Natural processes remain irreducibly teleological while remaining material processes. Aristotelians, then, agree with modern biologists on the importance of the underlying materials and mechanisms of organisms without at the same time accepting the reduction of the organism's reality to its material and biochemical mechanisms.[50]

Although Aristotle did not completely reject the mechanistic world-view postulated by Empedocles, he insisted that it could not provide a complete explanation of organic reality. As he says in *Parts of Animals*, "Even Empedocles ... finds himself constrained to speak of the reason (*logos*) as constituting the essence and real nature of things."[51] The necessity of recognizing both teleology and mechanism has been elegantly stated by the celebrated polymath, D'Arcy Wentworth Thompson: "Still, all the while, like warp and woof, mechanism and teleology are interwoven together, and we must not cleave to the one nor despise the other; for their union is rooted in the very nature of totality."[52] The interaction between mechanism and teleology also relates to contingency. As Etienne Gilson remarks, mechanism can only offer chance as an explanation, yet chance is not a cause but simply an absence of explanation. Opposing the neo-Darwinian notion that the whole evolutionary process is driven by chance events such as random copying errors in DNA, Gilson contends that "whether the absence of a cause lasts a year or billions of years, it remains forever an absence of cause, which, as such, can neither produce nor explain anything."[53]

It is a matter of fact that certain biological phenomena cannot be explained by mechanistic causation alone. A number of such cases have been pointed out by the Catholic theologian Terence Nichols,

50. Tkacz, Michael W., "Thomistic Reflections on Teleology and Contemporary Biological Research," in *New Blackfriars* 94 (2013): 671–72.
51. Translation in Gilson, *Aristotle*, 125–26.
52. Thompson, *Growth*, 5.
53. Gilson, *Aristotle*, 154.

including morphogenesis, or the development of (physical) form in organisms, the regeneration of damaged organs in animals, and the ability of various organisms to regenerate themselves from parts, such as the ability of flatworms to develop into complete animals when cut into pieces. Nichols asserts that none of these cases are explicable if formal causality is omitted.[54]

Aristotle acknowledges the role of mechanism in the development of living things, but, as stated earlier, holds the final cause to be primary. He writes in *Parts of Animals*: "Plainly, however, that cause is the first which we call the final one. For this is the Reason (*logos*), and the Reason forms the starting-point, alike in the works of art and in works of nature.... Now in the works of nature the good end and the final cause is still more dominant than in works of art such as these, nor is necessity a factor with the same significance in them all" (I.639b). Arguing for the primacy of final causality, Aristotle employs the analogy of a physician or a builder. The proper order of enquiry is not to start with the process of formation of each animal, but to consider first its actual characteristics and then deal with how they evolved.[55] As he states in the *Parts of Animals*, "the process of evolution is for the sake of the thing finally evolved, and not this for the sake of the process" (I.640a). The presence of the final cause is a necessary condition for the actualization of potential.

The reluctance among modern biologists to recognize final causality in nature, not to mention the outright hostility thereto, has been brilliantly satirized by the British polymath J.B.S. Haldane: "Teleology is like a mistress to the biologist: he cannot live without her, but he's unwilling to be seen with her in public."[56] Judging by the neo-Darwinist rejection of final causality, the "ghost of teleology" evidently continues to haunt those unwilling to admit a transcendent influence in the processes of life.

The Neoplatonists

Following Plato, Plotinus used the term "living being" (*zōion*) for

---

54. O'Rourke, *Metaphysics*, 30.
55. Ross, *Aristotle*, 128.
56. Quoted in O'Rourke, *Metaphysics*, 20.

Hellenic Philosophy of Life, or Bio-Philosophy

the composite of body and soul in an individual: "What we should say is that the living being is either a certain kind of body [living], or the sum [of body and soul], or some other third thing that arises from both of these" (*Enneads* I.1.5). Plotinus argues for this last view, that a living being is the product that arises from the combination of body and soul, and not merely the sum thereof (*Enneads* I.1.7, 1–6).[57] He argues further that just as the Intellect and the World-soul arise through contemplation (*theōria*, literally "a looking at, viewing, beholding, or observing"), so do the things of nature. He writes:

> That which comes to be is my vision, my act of silence, a thing contemplated that comes to be by nature, and since I come to be by contemplation that is like this, it is the case that I have the nature of a lover of contemplation. And my contemplating makes the product of contemplation, just as geometricians draw what they are thinking. But with me, I do not draw; rather, I contemplate, and the outlines of bodies materialize as if they resulted from my contemplation. (*Enneads* III.8.4)

This passage on the contemplative origin of physical bodies is a beautiful personification of nature.[58] Plotinus continues in the same passage, "And there exists in me my mother's state and the beings that generated me. Those, too, come from contemplation, and my becoming was through no action of theirs; rather, those greater expressed principles contemplated themselves, and I came to be." In this context "mother" refers to the World-soul, whereas the "beings that generated me" indicates the *logoi* in the soul derived from the Forms contained in the Intellect, according to which nature is constituted.[59]

Proclus held that the relation between the One and the many also pertains to the distinction between a genus and its species. Of the genus, he writes that it "is the 'unique form' spread through many separate things and existing in each of them," and adds that the genus pre-exists each of its species and is participated in by them

---

57. Dillon and Gerson, *Philosophy*, 278.
58. LSJ, 817; Theodossiou et al., *Theories*, 97.
59. Dillon and Gerson, *Philosophy*, 39.

and by the genus itself. He also states clearly that "the realities existing prior to species are not identical with the characters that exist in the species by participation" (*Commentary on Parmenides* I.5). It is therefore incorrect to view a genus as a "whole of parts"; rather, "the species are the many forms different from one another but comprehended by one unique embracing form, which is the genus."[60]

If we apply Proclus' metaphysical reasoning to the biological realm, we might say that species "evolve" out of their genus through participating in the pre-existent realities contained in the genus. For example, the magnificent big cats such as lions, tigers, leopards, jaguars and snow leopards partake as species in the genus *Panthera*. By extrapolation, we can also apply this to genera participating in their family (i.e., big cats in *Felidae*), to families in their order (i.e., cats in *Carnivora*), to orders in their class (i.e., carnivores in *Mammalia*), and to classes in their phylum (i.e., mammals in *Chordata*). Viewed the other way round, each phylum contains the causes of its classes, each class of its orders, each order of its families, each family of its genera, and each genus of its species.

The Great Chain of Being

The metaphysical notion of graded continuity between various levels of being, all of it originating in the divine Principle (i.e., the One), is often referred to as the Great Chain of Being. In the Western world the notion of levels of being is first encountered in the work of Homer, who mentions a Golden Chain (*seire chruseie*) stretching from heaven to earth (*Iliad* VIII.18).[61] The chain of being was applied for the first time to the organic realm in Aristotle's scale of nature that ranges from the Prime Mover downwards to unformed matter. The Neoplatonists elaborated the teachings of their Hellenic predecessors to postulate an all-embracing hierarchy of existent things ranging from inanimate matter through plants, animals, humans and heavenly beings (variously referred to as gods or angels) to God at the summit. It is to this Divine summit that

---

60. Ibid., 316.
61. Uzdavinys, Algis, *The Golden Chain. An Anthology of Pythagorean and Platonist Philosophy* (Bloomington, IN: World Wisdom Books, 2004), xxi.

Hellenic Philosophy of Life, or Bio-Philosophy

humans are called to return. The Sufi scholar S.H. Nasr, writes: "The 'great chain of being' was formulated not to explain away the Divine Cause but to enable man to ascend towards the Source while being fully aware of the total cosmic equilibrium of which natural forms provide such striking evidence."[62]

This all-embracing, hierarchical ontology became influential in both the Christian and Islamic worlds of the so-called Middle Ages.[63] Working at the Carolingian court in the ninth century, John Scottus Eriugena described the whole of reality from God to matter in his *Periphyseon*. The Irish scholar argued that God creates the cosmos through a five-fold process, out of pre-ontological nothingness (Greek *ouk on*, Latin *nihil*) through the primordial causes (Latin *causae primordiales*) similar to the Platonic Ideas, which are contained in the divine Logos. This creative activity results in the establishment of various levels of being: natural bodies, rocks, for example, are created for existence only; plants for existence as well as life; irrational animals for existence, life and sensation; human beings for existence, life, sensation and reason; and finally, heavenly beings (for example, angels) for existence, life, sensation, reason and intellect (*Periphyseon* II, 580).[64] In this way each level of being obtains its own characteristic element from the Logos, while also sharing in those below it.

In turn, Islamic scholars such as the Ikhwan al-Safa or Brethren of Purity and Ibn Sina, known in the Latin West as Avicenna, depicted the three kingdoms of minerals, plants and animals according to their degrees of perfection. Thus, Nasr writes, "the mineral world

---

62. Nasr, Seyyed Hossein, *Islamic Science. An Illustrated Study* (London: World of Islam Festival Publishing Company Ltd, 1976), 71.

63. We say so-called, since according to the Organic cultural model postulated by Oswald Spengler, Francis Parker Yockey and other thinkers, each High Culture, whether Indian, Chinese, Classical, Arabic, or Western, passes through its own phases of birth, growth, maturity, old age and death. Notions such as "ancient," "medieval" and "modern" are therefore quite relative, in addition to ideologically serving an Occidental view of history. However, for the sake of convention we will continue to use the adjective "modern" when referring to the Western world of the past four to five centuries, in which sense it also means "anti-traditional."

64. Carabine, Deidre, *John Scottus Eriugena* (New York and Oxford: Oxford University Press, 2000), 56.

ranges from the most opaque substances to those which resemble plant life, and plant life likewise stretches from moss and algae, which resemble mineral growth, to palm trees, in which there is a differentiation of the sexes. The same holds true for the animal kingdom, at whose apex are creatures which resemble man in certain aspects of their intelligence." We should note here that in Islamic philosophy, this gradation of beings was based not so much on anatomical resemblance as on the reflection of cosmic qualities.[65] We therefore suggest that animals possessing undoubted nobility such as lions and the other big cats, eagles and many other birds of prey, not to mention the blue whale and the polar bear, rank higher in the chain of being than do many, if not most members of the primate family.

In his *Summa contra Gentiles*, Thomas Aquinas uses the doctrine of divine providence to account for the ordered variety of the world. Since all created things fall short of the full goodness of God, there has to be variety in things in order to express perfection. Thomas writes that

> for the perfect goodness that exists one and unbroken in God can exist in creatures only in a multitude of fragmented ways. Now variety in things comes from the different forms determining their species. So because of their goal things differ in form. And because they differ in form there is an order among things. By dint of having form a thing exists, and by existing it resembles God, who is his own existence pure and simple; so form can be nothing else than God's resemblance in things. (3.97–98)[66]

Incidentally, here again is an instance of the metaphysical principle of differentiated unity, or the One-in-many. Thomas adds that this ordered variety of forms in nature harbors a number of implications. First, it involves levels of being, since forms differ by degree of perfection. Thus variety in things is achieved in steps, from inanimate matter through plants and unreasoning animals to intelligent

---

65. Nasr, *Science*, 71.
66. McDermott, *Aquinas*, 271.

creatures. Second, the diversity of form in different species results in different behavior, and therefore things differing in form act in different ways. Third, the variety of forms entails different relationships to matter. Whereas some forms can exist by themselves due to their degree of perfection, others need matter as a sort of base. And since matter and form cannot combine without some proportion between them, we find that different materials match different forms. These diverse relationships to matter produce a diversity of agencies, while the combined diversity of form, matter and agency brings about a diversity of characteristic and incidental properties (*Summa contra Gentiles*, 3.97–98).[67] In this way is the concept of a chain of being cast in terms of the Aristotelian metaphysics of form and matter.

Although the Western world was becoming increasingly secularized, the chain of being was still widely known as late as the eighteenth century. As a matter of fact, the Hellenic principles of plenitude, continuity and gradation found their widest acceptance at precisely this time.[68] A celebrated example of this can be seen in the work of Carl Linnaeus, the Swedish botanist who made use of the chain of being in his binomial system of biological classification, published as *Systema Naturae* in 1735. This work assigned both a generic and a specific name to thousands of animal and plant species, and it still forms the basis of modern taxonomy. However, Linnaeus diverged from the traditional view by postulating a class of primates, consisting not only of monkeys and apes but also human beings.[69] It is worth noting that Linnaeus drew inspiration from the Biblical account of Adam naming all the creatures in the Garden of Eden according to God's command, and his two-name system applied uniformly across the animal, plant, and mineral kingdoms "reflect[s] the idea that Adam's exercise was informed by a *logos*

---

67. Ibid., 271–72.
68. Lovejoy, Arthur O., *The Great Chain of Being. A Study of the History of an Idea* (New York: Harper & Brothers, 1960), 183.
69. Van Vrekhem, Georges, *Evolution, Religion and the Unknown God* (North Charleston, SC: CreateSpace Independent Publishing Platform, 2012), 16–17.

emanating from God that identified a function corresponding to each species as part of an overall plan."[70]

However, during the eighteenth century the chain of being also became temporalized in much of European thought. The principle of plenitude came to be viewed temporally as "a process of development in time, materialized in the gradually increasing complexity of the organisms on Earth."[71] It was this temporalized and materialized understanding of the chain of being that contributed to the rise of evolutionary thought in its modern, horizontal form. The divergence from an earlier view of evolution has been noted by Oswald Spengler. For the thinkers of the nineteenth century, evolution meant progress in the sense of increasing fitness of life towards purposes. In contrast, their predecessors such as Leibniz and Goethe understood evolution as "fulfillment in the sense of increasing connotation of the form."[72] This latter notion approaches the traditional understanding of evolution far more so than the increasingly horizontal ideas of the nineteenth century.

After the metaphysical tradition had been all but forgotten in the West due to the destructive rise of rationalism and materialism, as well as the burgeoning "progress" mythology, major aspects of the traditional Hellenic ontology were restated by Arthur Lovejoy in his *The Great Chain of Being*, first published in 1936. In this epochal work the author drew together three principles, or so-called "unit ideas," obtained from Plato and Aristotle: plenitude, continuity and gradation.[73] The principle of plenitude (from the Greek *plērēs*, meaning "full of," hence full, complete, or sufficient) comes from Plato, notably his dialogues *Politeia* (usually translated as *Republic*) and *Timaeus*. It implies that the wide range of kinds of living things is abundantly exemplified in the universe; that "no genuine poten-

70. Fuller, Steven, *Dissent over Descent. Intelligent Design's Challenge to Darwinism* (Cambridge: Icon Books, 2008), 63.
71. Lovejoy, *Chain*, 244; Van Vrekhem, *Evolution*, 95.
72. Spengler, Oswald, *The Decline of the West* Abridged edition, trans. Charles Atkinson, ed. Helmut Werner (Oxford: Oxford University Press, 1991), 231.
73. Bynum, William F., "The Great Chain of Being after forty years: an appraisal," in *History of Science* 13 (1975): 4. http://adsabs.harvard.edu/abs/1975HisSc..13....1B.

Hellenic Philosophy of Life, or Bio-Philosophy

tiality of being can remain unfulfilled"; and that the extent and abundance of life-forms must match the possibility of existence, commensurate with the productive capacity of an inexhaustible, transcendent Source. Nonetheless, Lovejoy cautions that this notion of plenitude should not be confused with perfection. "The principle of plenitude is rather the principle of the necessity of imperfection in all its possible degrees."[74] It is reasonable to think that Plato would have agreed with this caveat, given his insistence in the *Timaeus* that the rational works of the Demiurge are limited by the ubiquitous presence of irrational Necessity.

Although, as Lovejoy notes, Aristotle rejected the necessary actualization of all potentialities (*Metaphysics* III.1003a), he applied the principle of continuity to the organic realm when he recognized the existence of continuous transitions from the inanimate to the animate, and from plants to animals (*History of Animals* VIII. 588b). Lovejoy concludes that the Platonic principle of plenitude logically implies the principle of continuity, and writes, "If there is between two given natural species a theoretically possible intermediate type, that type must be realized..., otherwise, there would be gaps in the universe, the creation would not be as 'full' as it might be."[75] The third Hellenic "unit idea" is the principle of gradation, described by Aristotle as a scale of nature based on the degree of perfection of beings ranging from zoophytes to humans (*Generation of Animals* II.732a–733b). By combining these principles of fullness, continuity and gradation, "The result was the conception of the plan and structure of the world... the conception of the universe as a 'Great Chain of Being,' composed of... an infinite number of links ranging in hierarchical order from the meagerest kind of existents... through 'every possible' grade up to the *ens perfectissimum*."[76]

Eventually, writes Lovejoy, these Platonic and Aristotelian ideas were organized into a coherent system in the Neoplatonic theory of emanation. In the *Enneads* Plotinus says that the Soul generates a reflection of itself in sense-perception and the principle of growth,

---

74. LSJ, 565; Lovejoy, *Chain*, 52, 338.
75. Lovejoy, *Chain*, 55, 58.
76. Ibid., 59.

and that the organic realm comes to be as a result of this (V.2.1). The organic realm appears "like a long line stretched out in length, each of the parts other than those that come next, all continuous with itself, but one being different from the other, the first not being destroyed with [the appearance of] the second" (V.2.2).[77] And since the generation of these levels of being (i.e., plants and animals) is a logical necessity, Plotinus regards phenomena such as inequality, imperfection, and suffering as necessary for the good of the whole.[78] Plotinus denies that the existence of inequalities is due to the divine Will, as is taught in both Christian and Islamic theology, instead asserting it as necessary according to the nature of things (*Enneads* III.3.3).

It is notable that even after the apparent victory of Neo-Darwinism over alternative theories of evolution, in the middle of the twentieth century, the chain of being continued to inspire some Western thinkers. Prominent among these was the German economist and statistician, Ernst Schumacher, who, in his *A Guide for the Perplexed* (1977), depicted an ascending hierarchical order of minerals, plants, animals, and humans. Parallel to these four "kingdoms" are four characteristic elements, namely, matter, life, consciousness, and self-awareness. Schumacher held that each ontological level, or kingdom, possesses its own characteristic element while also sharing those of the levels below, recalling Eriugena's similar schema outlined above. Schumacher further maintained that the differences between the levels of being represent fundamental ontological discontinuities. The science of physics properly deals with the lowest level only, i.e., matter; philosophy is outside its area of competence.[79] This agrees with Plato's conviction that natural science deals with a lower level of knowledge than philosophy, since its domain is the lower, sensible world which exists through participation in the higher, intelligible world. Consequently, knowledge of the material world is only opinion and not true knowledge, which is limited to the intelligible world.

---

77. Lovejoy, *Chain*, 59, 61; Dillon and Gerson, *Philosophy*, 85–86.
78. Lovejoy, *Chain*, 64–65.
79. Van Vrekhem, *Evolution*, 97–98.

Hellenic Philosophy of Life, or Bio-Philosophy

Paul Davies is a contemporary scientist who has defended the notion of a structural hierarchy in nature that cannot be explained in material terms only. In his *God and the New Physics* (1984), the English physicist argues that realities such as life, organization and mind cannot be reduced to the actions of atoms, quarks or unified forces. In an admirable disregard of the prescribed divorce of physics from metaphysics in the modern scientific project, Davies writes that "life is a holistic concept, the reductionist perspective revealing only inanimate atoms within us. Similarly mind is a holistic concept, at the next level of description. We can no more understand mind by reference to brain cells than we can understand cells by reference to their atomic constituents."[80] In other words, the ontological discontinuities from matter to life and from life to mind cannot be explained by reducing these discontinuities to material and mechanical factors, as much of modern science has vainly tried to do.

Here we must be clear that the foregoing affirmation of a hierarchical ontology implicitly denies the theological notion of a "God of the gaps." According to the argument, popular among some religious believers who—often rightly—reject the dogmatic assertions made by atheist scientists, the created order is characterized by developmental "gaps" which, to bridge, require miraculous interventions by God as Creator, thus assuring "both God's existence and omnipotence."[81] Gaps in scientific knowledge, as well, are assumed to provide evidence of divine intervention. For instance, advocates of the Intelligent Design movement often invoke a transcendent Designer as an *immediate* explanation for the existence of highly complex organic systems, rather that recognizing such a Designer as an *ultimate* explanation, with an unknown number of layers in between.[82] This is consonant with our earlier discussion of the distinction between the First Cause of the universe and the mediate causes through which the physical world, including the living kingdoms, come into existence.

---

80. Quoted in Van Vrekhem, *Evolution*, 100.
81. Van Till, *Integrity*.
82. Marshall, *Evolution*, 212–13.

Finally, we ought to note that the metaphysical notion of a cosmic chain of being does not necessarily imply a single, absolute hierarchical system. Applying insights obtained primarily from the metaphysician Frithjof Schuon, James Cutsinger takes a broader, multi-perspective approach, when he says that "while plants may in general be 'higher' than minerals, because of the organic life that is in them, a precious gem remains a more lucid theophany than a weed. The same qualification must be applied to the relative positions in the great chain of being of plants and animals, and of animals and man. A noble animal, like an eagle or a lion, is 'more Divine' than a human being who lives below himself."[83] We are also reminded hereby of the Traditionalist position just mentioned, that some animal species stand closer to the human being in their manifestation of Divine qualities, such as nobility, than is the case with many members of the primate family.

Cutsinger's reference to humans living below themselves is reminiscent of the words Friedrich Nietzsche puts into the mouth of his hero at the beginning of *Thus Spoke Zarathustra*. Zarathustra makes his first speech in the market-place after descending from his mountain abode: "You have made your way from worm to man, and much in you is still worm. Once you were apes, and even now man is more of an ape than any ape."[84] By living below themselves, humans are failing in their divine calling to become like God through an increasing participation in the Divine energies, as taught by the Greek Patristic theologians. In the words of the fourth-century Basil, Archbishop of Caesarea, "God created man like an animal who has received the order to become God."[85] Situated midway between the heavenly beings and the animals on the chain of being, human beings have the freedom to choose whether to orientate themselves upward towards God or downward towards the apes.

---

83. Cutsinger, *Earth*, 24.

84. Nietzsche, Friedrich Wilhelm, *Thus Spoke Zarathustra*, trans. R.J. Hollingdale (Harmondsworth: Penguin Books, 1969), 42.

85. Lossky, Vladimir, *Orthodox Theology. An Introduction*, trans. Ian & Ihita Kesarcodi-Watson (Crestwood, NY: St Vladimir's Seminary Press, 1978), 73.

4

Form and Transformation

IN AN EARLIER CHAPTER we discussed the numerical and geometrical foundations of the cosmos as outlined by Pythagoras, Plato, and Iamblichus. The process of cosmic manifestation entails an "unfolding"—which is the precise meaning of "evolution"—from numbers into lines (i.e., the point becoming one-dimensional), from lines into planes (i.e., the line becoming two-dimensional), and from planes into solids (i.e., the plane becoming three-dimensional). That is to say, the movement from the One into the many proceeds in the following sequence: numbers → lines → geometrical figures → solid bodies, according to the patterns of the eternal Forms. Moreover, as we have argued, the phenomenon of organic form (*morphē*) cannot be conceived in material terms alone, since the Forms exist prior to their formation of material objects (Plato), which in the case of living beings entails the priority of soul over body (Aristotle).

As we observed above, Pythagoras and Plato held the order of nature to be mathematical, or, more precisely, geometrical, viz. Plato's cosmological account in the *Timaeus*. Indeed, Plutarch attributes to Plato the phrase, "God is always geometrizing" (*ton theon aei geōmetrein*). The paradigmatic significance of geometry for Plato is further evident from the inscription above the entrance to his Academy: "Let no one ignorant of geometry enter here." The biochemist Michael Denton notes that since the laws of nature can be described in mathematical forms, nature is inherently intelligible. Aristotle also recognized the correspondence between the human mind and the natural world, for example his association in the *Metaphysics* of the sense of sight with our knowledge of the world.[1]

Earlier, in our discussion of Hellenic cosmology, we introduced the Pythagorean sacred number ten, the *tetraktys*. The metaphysical and physical significance of the *tetraktys*, that is, 1+2+3+4, has been sketched as follows: The One, or the monad, is "the limit of all, first before the beginning and last after the end, *alpha* and *omega*, the mold that shapes all things and the one thing shaped by all molds, the origin from which the universe emerges, the universe itself, and the center to which it returns. It is point, seed, and destination."[2] We should keep in mind that these terms are symbolic, since in reality the One is an ineffable mystery, as is emphasized in both the Hellenic and Patristic traditions. Two is the dyad, viewed by the Pythagoreans as an even and female number, and serves as the basis of all the pairs of opposites in the cosmos, such as limited-unlimited, one-many, and male-female. The dyad also appears in music as the ratio 2:1, or the octave, and as the Sun and the Moon under which we live by day and by night. Three is the triad that serves as mediator between opposites, thus returning them to unity. The triad manifests in time as past, present, and future; in geometry as the triangle, which is the first polygon; and in the musical ratios 3:2 and 3:1, i.e., the fifth and its octave. Four is the tetrad, symbolizing the Earth and the natural world. The tetrad appears in geometry as the simplest solid figure, the tetrahedron; in manifestation as the material elements of fire, air, earth, and water; in music as the ratios 4:3 and 4:1, the fourth and two octaves; and in the particles of which all material things consist, namely protons, neutrons, electrons, and neutrinos.[3] From the foregoing, we see that the manifestation of the first four numbers or *tetraktys* undergirds the realm of inorganic form.

Organic Form

In the traditional Hellenic understanding, the number five, or the

---

1. Denton, Michael, *Nature's Destiny. How the Laws of Biology Reveal Purpose in the Universe* (New York: The Free Press, 1998), 240, 392.
2. Lundy, *Number*, 12.
3. Ibid., 14–18.

pentad, is associated with life (*bios*), and is also the number of water. In her book on sacred number, Miranda Lundy writes: "Water itself is an amazing liquid crystal lattice of flexing icosahedra, these being one of the five Platonic solids, five triangles meeting at each point. As such, water shows its quality as being that of flow, dynamism, and life."[4] The Neoplatonist philosopher Iamblichus held that the lowest level of organic life is that of plants, of which the general structure is five-fold: root, stem, bark, leaf, and fruit. While the first four numbers account for the generation of three-dimensional bodies, as we have noted, the pentad entails change in respect of addition and increase, which is organic growth. It thus represents the vegetative aspect of the soul. The five sense-organs found in the higher animals are related to the five general elements, including ether.[5] Geometrically, the number five is represented by the pentagram, which is an image of the human being. It also indicates the elements of earth, fire, air, and water as modalities of the central element, ether, and it signifies the mental faculties of reason, intuition, imagination, and memory as emanating from the Intellect. In some traditions this notion is visually expressed by the five-petalled rose in the center of the cross, symbolizing the unmanifested quintessence (Latin *quinta essentia*) which is the central principle of the four manifested elements.[6]

As for six, or the hexad, it occurs in crystalline structures such as snowflakes, quartz, and graphite, in the hexagons of carbon atoms that form the basis of all organic chemistry, in the hexapodal locomotion of insects, and in the hexagonal honeycombs of bees. D'Arcy Wentworth Thompson offers a Platonic interpretation of the forms of snowflakes or snow-crystals when he writes, "These owe their multitudinous variety to symmetrical repetitions of one simple crystalline form—a beautiful illustration of Plato's One among the Many."[7] This again brings to mind the world as described by tradi-

---

4. Ibid., 20.
5. Iamblichus, *Theology*, 67–68, 73.
6. Schuon, Frithjof, *Esoterism as Principle and as Way*, trans. Willian Stoddart (Middlesex: Perennial Books, 1981), 74; Northbourne, *Progress*, 97.
7. Lundy, *Number*, 22; Thompson, *Growth*, 153.

tional cosmology, where it is a differentiated unity, the many receiving their being from the One.

The number seven, the heptad, appears in the colors of the rainbow; in the days of the week, of which each is connected with a specific planet and metal; and in the groups of crystal structures. The first cubic number after one, eight (i.e., the octad, as 2 x 2 x 2), is manifested geometrically as the octagon, which indicates the link between Heaven, symbolized by the circle, and Earth, symbolized by the cube. The spatial perfection of the number eight is displayed in nuclear physics, where atoms "desire," as it were, to have a full octave of eight electrons in their outermost shell, e.g., a sulphur atom has six electrons in its outermost shell, and therefore eight atoms join to share electrons, thus forming an octagonal sulphur ring. It was further remarked by Iamblichus that in the organic realm the octad is manifested in the number of feet of arachnids and crustaceans (although the Neoplatonist erred regarding the latter, which are actually decapods that use the first pair of feet as pincers); in the arrangement of human teeth as four quadrants of eight teeth each; and in the apertures of the mammalian head, namely pairs of eyes, ears and nostrils, and the two channels for air and food, respectively, in the mouth.[8]

The number nine, the ennead, is the celestial number of order, being the square of the triad. It appears as the sum of the regular three-dimensional shapes, namely the five Platonic solids and the four stellar Kepler-Poinsot polyhedra; the nine planets of our solar system; the cross-section of the tentacle-like cilia on our internal surfaces; and the bundles of microtubes in centrioles, which are essential for cell division. As for the next number, ten, the decagon, it is noteworthy that geometrically it is formed from two pentagons. And since five is the number associated with water and organic life, it is highly significant that DNA, the biochemical key to the reproduction of life, has ten steps for each turn of its double helix and appears in cross-section as a tenfold rosette.[9]

We have already noted the relation between the numbers 5, 8, and

8. Lundy, *Number*, 24–26; Iamblichus, *Theology*, 102.
9. Lundy, *Number*, 28–30.

13 as encountered in two- and three-dimensional geometry. These numbers also occur in the Fibonacci sequence of 1, 1, 2, 3, 5, 8, 13, 21, 34, 55, and so forth, in which each number from 2 onwards is the sum of the preceding two numbers. We find biological examples of the Fibonacci sequence in the arrangement of leaves on a stem, the fruitlets of a pineapple, the flowering of an artichoke, an uncurling fern, the arrangement of a pine cone, and the genealogy of honeybees. The latter is due to the fact that male bees have only one parent, the queen, since they hatch from unfertilized eggs, whereas female bees have two parents. The family tree of any male bee will therefore show 1, 2, 3, 5, 8, 13,... ancestors going back through its generations.[10]

It is noteworthy that most of the plants on Earth produce alternate leaves at Fibonacci fractions of a full rotation. This phenomenon is called phyllotaxis, from the Greek *phyllas* (leaf) and *taxis* (rank or order). As explained by John Martineau, "Some plants produce leaves along a stem every 1/2 rotation, in hazel and beech trees the angle is 1/3, in apricot and oak trees it is 2/5, in pear and poplar trees it is 3/8, and in almond and willow trees it is 5/13." In addition, "pineapples display 5-, 8-, and 13-armed spirals (viewed from near-horizontal, at 45° and vertical, respectively), and a willow sprig displays 13 buds in 5 turns. This striking phenomenon not only follows mathematical laws but also serves a definite purpose, optimizing the access of plants to sunlight and dew."[11] This demonstrates once again that organic functionality is optimized by conformity of the sensible form (*morphē*) to the intelligible form (*eidos*).

In the Golden Ratio, we find another instance of mathematical law applied to the organic realm. Two quantities are in this ratio if the sum of the two quantities divided by the larger of the two has the numerical value of approximately 1.618 (more precisely, 1.618033988...). This ratio has been studied by mathematicians since at least the time of Euclid in the third century BC. In the nineteenth century, Adolf Zeising claimed to have found the Golden Ratio in the arrangement of the parts of plants, the skeletons of animals, the

10. https://en.wikipedia.org/wiki/Fibonacci_number#In_nature.
11. Martineau, John, "A Little Book of Coincidence," in *Quadrivium*, 324–25.

proportions of chemical compounds, and the geometry of crystals. He concluded that the Golden Ratio functions as a universal law throughout the inorganic and organic kingdoms, culminating in the human form. More recently, scientists such as Jean-Claude Perez discovered the Golden Ratio in the human genome, with the percentages of 64 different codons following a fractal pattern based on the ratio 1.618.[12]

We encounter a further instance of mathematical regulation in the plant kingdom with botanical shapes that can be accurately depicted in terms of sine curves. These latter, also known as sine waves, are used to describe smooth, repetitive oscillations. The sine curve can be found among the following types of plants: (a) reniform or kidney-shaped leaves, as in the ground-ivy and dog-violet; (b) pentamerous or five-petalled flowers; and (c) composites of (a) and (b), as in the horse-chestnut leaf with its five petal-like leaflets.[13]

Applying the principles of both Hellenic metaphysics and Newtonian physics to the organic realm, D'Arcy Wentworth Thompson argued in his work *On Growth and Form*[14] that both the forms of natural phenomena (sea, clouds, etc.) and the material forms of living things (cell and tissue, shell and bone, leaf and flower, etc.) obey the laws of physics. Thus, Thompson writes, the problems of forms are primarily mathematical problems, while their problems of growth are essentially physical problems.[15] Morphology consequently becomes interwoven with mathematics and physical science.

Thompson introduces his argument by emphasizing the significance of number in the investigation of nature, while giving a nod to Pythagoras and Plato. He writes:

---

12. https://en.wikipedia.org/wiki/Golden_ratio; Marshall, *Evolution*, 304.
13. Thompson, *Growth*, 282–84.
14. In this work, written during the First World War and expanded during the Second, the author's magisterial grasp of not only biology but also the classics and mathematics is evident from the seamless integration of these seemingly disparate disciplines into a tome that has been referred to as the finest work of scientific literature in the English language (Stephen Jay Gould, Foreword to the Abridged Edition, 1992).
15. Thompson, *Growth*, 7–8, 269.

Form and Transformation

> As soon as we adventure on the paths of the physicist, we learn to *weigh* and to *measure*, to deal with time and space and mass and their related concepts, and to find more and more our knowledge expressed... through the concept of number, as in the dreams and visions of Plato and Pythagoras; for modern chemistry would have gladdened the hearts of those great philosophical dreamers. Dreams apart, numerical precision is the very soul of science.

Thompson also affirms the importance of final causality in the investigation of the living world. Along with Aristotle, he holds that "the organism is the *telos*, or final cause, of its own processes of generation and development."[16] The reader is then guided through hundreds of pages devoted to numerous instances of mathematical lawfulness as observed, for example, in the forms of cells and tissues, including honey-combs and blood vessels; the spicules of sponges and the skeletons of radiolara; the spiral forms of mollusc and gastropod shells; the shapes of horns, teeth and tusks; and the structure of bones and of whole skeletons, with these being explained in terms of forces such as tension, compression, stress and strain.

We should note that throughout this epochal work Thompson appears to conceive "form" in the sense of *morphē* (sensible shape) rather than *eidos* (intelligible structure).[17] Nonetheless, the Hellenic notion of cosmic lawfulness is affirmed by Thompson when he states that in general no organic forms exist without conformity to physical and mathematical laws. He says, in addition, that the phenomenon of growth should be studied in relation to form, whether this growth is understood as increase in size or as gradual change of form. This recognition of physical and mathematical laws in nature does not entail a static view of the organic world, but rather a dynamic one. The form of any portion of matter, and the changes of form that appear in its movements and growth, are due to the actions of force. Thompson writes that in the case of organisms the nature of motions is interpreted in terms of force, i.e., kinetics, whereas the conformation of the organism itself is interpreted in terms of the balance of forces, i.e., statics. Morphology deals not

---

16. Ibid., 1–4.
17. Bakar, *Criticism*, 176.

only with the study of material things and their forms, but also with the operations of energy, or dynamics.[18]

Thompson asserts that variation and inheritance cannot be reduced to material causes: "Matter as such produces nothing, changes nothing, does nothing; and ... we must most carefully realize in the outset that the spermatozoon, the nucleus, the chromosomes or the germ-plasma can never *act* as matter alone, but only as seats of energy and as centers of force." In this way he expresses in modern scientific terms the Hellenic philosophical dictum "for nature is very much the beginning of matter" (*arkhē gar he physis mallon tes hylēs*).[19] It also recalls the Hellenic doctrine regarding the priority of form over matter, according to which unformed matter exists as a pure potentiality until its actualization by form.

Theory of Transformations

Morphology is the study of organic form. It is relatively easy to pass from the mathematical concept of form in its static aspect to form in its dynamic relations, since these latter can be represented by a diagram of forces in equilibrium and the direction of forces that have effected the conversion of one form into another. Although not all organic forms can be described mathematically, enough instances exist to enable us to keep the various types in mind and leave the single, accidental cases alone. It is worth noting that in the mathematical Theory of Groups a distinction is made between substitution and transformation, the former being discontinuous and the latter continuous. As Thompson observes, this is curiously analogous to the biological distinction between mutation and variation.[20] In later chapters we will likewise argue that macro-evolution occurs through mutational "jumps," while micro-evolution takes place through minor variations.

The theory of transformations postulated by Thompson is based on the method of co-ordinates devised by the philosopher and mathematician René Descartes, which enables us to translate the

---

18. Thompson, *Growth*, 10–11, 14.
19. Ibid., 14.
20. Ibid., 269–71.

form of a curve into numbers and words. If a system of coordinates is deformed or submitted to a homogeneous strain, the result is a new figure that "is a function of the new coordinates in precisely the same way as the old figure was of the original coordinates x and y," as, for example, in cartography, when identical data is projected differently.[21] The morphologist takes the reverse course, not applying a new and artificial projection but rather trying to determine whether two different but more or less related organic forms can be shown to be transformed representations of each other, and, if so, postulating the direction and magnitude of forces capable of effecting such a transformation. Since the living body is an integral and indivisible whole, these co-ordinate diagrams should deal with organisms in their integral solidarity. If dissimilar fishes, for example, can be referred to identical functions of different co-ordinate systems, then it will constitute proof "that variation has proceeded on definite and orderly lines, that a comprehensive 'law of growth' has pervaded the whole structure in its entirety, and that some more or less simple and recognizable system of forces has been in control."[22]

The geometrical basis of Cartesian transformations is that a circle inscribed in a net of rectangular equidistant co-ordinates with horizontal axis X and vertical axis Y, can be elongated into an ellipse by decreasing the length of one axis. For example, the metacarpal or cannon-bone of an ox, a sheep, and a giraffe differ widely in size, but all are transformations of a basic type with an elongation along the vertical axis, in the ratio 3:2:1. Radial co-ordinates are applicable especially to organisms of which the growing structure includes a "node," a point where growth is absent or at a minimum, while around it the rate of growth increases symmetrically. This phenomenon occurs notably in the leaves of dicotyledon plants, flowering plants in which the seed has two embryonic leaves, or cotyledons.[23]

Thompson presents a number of special cases of organic transformation according to mathematical laws. One such striking instance is the transformation of a circle or sphere into two circles or spheres

21. Ibid., 272.
22. Ibid., 271–72, 275.
23. Ibid., 276–78, 280; https://en.wikipedia.org/wiki/Dicotyledon.

in the case of a small round gourd that, unimpeded, grows into a large round or oval pumpkin or melon. However, if a rag is tied around the middle of the gourd, it grows into two connected globes. Thompson writes, "It is clear, I think, that we may account for many ordinary biological processes of development or transformation of form by the existence of trammels or lines of constraint, which limit and determine the action of the expansive forces of growth that would otherwise be uniform and symmetrical."[24] These constraints on organic transformation serve as evidence in favor of directed evolution, or evolution within limits.

In his 1861 novel *Elsie Venner*, the physician and writer Oliver Wendell Holmes describes an interesting parallel between the operations of a glass-blower and those of nature. Holmes writes that both nature and the glass-blower start with a simple tube. Thus in nature the alimentary canal, the arterial system (including the heart), and the central nervous system of vertebrae (including the brain), all begin as simple tubular structures. Thompson explains the parallelism as follows: "And with them Nature does just what the glass-blower does, and, we might even say, no more than he. For she can expand the tube here and narrow it there; thicken its walls or thin them; blow off a lateral off-shoot or caecal diverticulum; bend the tube, or twist and coil it; and infold or crimp its walls as, so to speak, she pleases." Moreover, Thompson remarks, the handiwork of the glass-blower is an example of mathematical beauty.[25]

Variations Based on Rectangular Co-Ordinates
In his works *Geometry* and *Treatise on Proportion*, the famed Renaissance artist Albrecht Dürer described the transformation of human physical features by means of slight variations in the relative magnitude of the component parts, based on rectangular co-ordinates. Dürer's illustrations inspired the eighteenth-century physician and comparative anatomist Peter Camper's notion of facial angles, the variety of which can be represented by both rectangular and oblique co-ordinates, as had indeed been done by Dürer. This

---

24. Thompson, *Growth*, 286–87.
25. Ibid., 287.

geometrical depiction of human facial angles recognizes the essential fact that the skull varies as a whole, and that the facial angle is indeed the index to a general deformation.[26]

Surveying transformation among the Crustacea, Thompson demonstrates that the ostensibly immense difference between the elongated body of *Oithona nana,* an abundant zooplankton species, and the thick-set body of the genus *Sapphirina* or sea sapphires, entails nothing more than a difference of relative magnitudes, which can be expressed by means of rectilinear co-ordinates. In the case of crabs, the large variety of shapes of the carapace can all be represented by means of either equidistant rectangular co-ordinates and its elongation and compression, or curvilinear triangular diagrams. The same method applies to a variety of amphipods belonging to different families, for example the genera *Harpinia, Stegocephalus* and *Hyperia.* The mathematical basis of organic transformations does not only pertain to larger invertebrates but can be found among the hydroid zoophytes, a polymorphic group with apparently infinite variations in form, size, and the arrangement of the little cups at their top called calycles. Even these minute variations are a case of relative magnitudes that can be represented by means of Cartesian co-ordinates.[27]

Cartesian co-ordinates can again be used to represent the great variety of deformations found among fish. Thompson offers the following examples: A) If we take the rectangular co-ordinates that have been applied to an outline of the hatchet-fish (*Argyropelecus olfersi*) and shear them to oblique co-ordinates with angles of 70% and then apply the original outline to these new coordinates, the result is an outline of the allied diaphanous hatchet-fish (*Sternoptyx diaphana*). This deformation is analogous to the simplest and most common deformation of fossils due to the shearing-stresses in solid rock. B) If the rectangular co-ordinates applied to the parrotfish (genus *Scarus*) are deformed into a system of co-axial circles, the original outline then becomes the outline of the allied angelfish of the genus *Pomacanthus*, which fish—and this is most significant—

26. Ibid., 290–92.
27. Ibid., 292–97.

displays colored bands on its body corresponding to lines of curved ordinates. C) We can derive the outline of the short bigeye (*Pseudopriacanthus altus*) and the scorpionfish (genus *Scorpaena*) from that of the wreckfish (genus *Polyprion*), if we replace the rectangular coordinates of the wreckfish outline with a triangular or radial system of coordinates. D) The rectangular co-ordinates of the outlined porcupine-fish or balloonfish (genus *Diodon*), if deformed into a combination of concentric circles and hyperbolic curves, provide an accurate outline of the closely allied though very different-looking sunfish (*Mola mola*) if the original outline is transferred to the new coordinates. In this last case, the Cartesian method accounts by one integral transformation for all of the external differences between the two fish. Thus it leaves the parts near the origin of the system, such as the head and the pectoral fin, practically unchanged, and shows an increasing modification of size and form when moving from the origin towards the periphery of the system.[28]

Regarding the class Reptilia, the order of crocodiles is of particular interest to the study of evolution in that it provides us with an almost unbroken series of transitional forms in continuous succession across geological formations, as was pointed out by the Victorian biologist, Thomas Huxley.[29] A system of Cartesian co-ordinates and its subsequent deformations can be used to plot the transitions in a variety of crocodile skulls, for example that of the saltwater crocodile (*Crocodylus porosus*), the American crocodile (*Crocodylus americanus*) and the small Cretaceous form *Notosuchus terrestris*. The same method can also be applied to a variety of Dinosaurian reptiles. For instance, the medium-sized pterosaur *Dimorphodon* can be plotted onto rectilinear Cartesian co-ordinates and related to the very large *Pteranodon*, with its elongated jaws and backwardly directed crest, by means of a system of oblique co-ordinates in which the parallel lines become diverging rays.[30]

The realm of mammalian skulls represents an especially fertile field for the application of Thompson's theory of mathematical

---

28. Ibid., 298–301.
29. Ibid., 301–02.
30. Ibid., 302–03, 305–06.

transformations. To begin with, the same method of Cartesian co-ordinates and its transformations is applied to extinct rhinoceros types, for example *Hyrachyus agrarius* and *Aceratherium tridactylum*. Comparing the skulls of *Hyrachyus* and *Aceratherium*, Thompson notes that "the long axis of the skull of *Aceratherium* has undergone a slight double curvature, while the upper parts of the skull have at the same time been subject to a vertical expansion, or to growth in somewhat greater proportion than the lower parts. Precisely the same changes, on a somewhat greater scale, give us the skull of an existing rhinoceros."[31] Almost incredibly, Thompson's system of transformation applies even from rhinoceros to rabbit skulls. Thus Cartesian co-ordinates are used for the *Hyrachyus* skull and with a uniform flexure in a downward direction for the rabbit skull. Moreover, the enlargement of the eye and a modification in the form and number of teeth from *Hyrachyus* to rabbit constitute independent variations outside of general transformations, but not contradictory to it.[32]

The same method can be applied to the horse and its ancestors. According to modern evolutionary theory, the horse family Equidae (with its 9 extant species in a single genus, *Equus*) descended from a dog-sized mammal that lived in North America around 54 million years ago. During its evolution the equids are said to have migrated to Asia and Africa, where the zebra and donkey later appeared. The skull of the smallest form, *Eohippus*, can be placed in a Cartesian network which is correspondingly enlarged towards the largest form, *Equus*. We find that in the case of intermediate forms such as *Mesohippus* and *Protohippus* the fossil skulls coincide with the hypothetical forms. Nonetheless, the skull of *Parahippus* does not fit into this scheme, Thompson observes, and therefore *Parahippus* cannot stand in the direct line of descent from *Eohippus* to *Equus* but has to represent a divergent branch of the Equidae.[33]

When comparing human skulls with the skulls of some of the higher apes, the main differences pertain to the enlargement of the

---

31. Ibid., 310–11.
32. Ibid., 313–14.
33. Cooke, *Animals*, 168; Thompson, *Growth*, 314–15.

brain and the relative diminution of the jaws, as well as the facial angle increasing from oblique to nearly a right angle in *Homo sapiens*. Thus the human skull is depicted with Cartesian co-ordinates, its transformation for the chimpanzee and a more intense deformation for the baboon. Consequently, "it becomes at one manifest that the modifications of jaws, brain-case, and the regions between are all portions of one continuous and integral process." However, Thompson correctly remarks, there exists no series of transitional forms between Mesopithecus, Pithecanthropus, *Homo neanderthalensis* and the various races of *Homo sapiens*. This lack of transitional forms between modern humankind and its alleged ancestors argues that no straight line of descent exists. Instead, among both human and anthropoid types we find divergent rather than continuous variation.[34] In addition, the main advantages of the human being, namely his brain and hands, would (at least initially) have been of little competitive advantage over his supposed ancestors with their greater strength and agility, as the British lawyer and ornithologist Douglas Dewar has pointed out.[35] The media-promoted speculation concerning a "missing link" between the higher apes and *Homo sapiens* therefore has to be rejected, even on strictly physical grounds, as an exercise in futility.

Any two mammalian skulls may be compared with this method, Thompson continues, since there is something invariant in spite of the transformations. Thus the landmarks of cranial anatomy, for example ear, eye and nostril, retain their relative order and position throughout a series of transformations. In addition, there exists a degree of invariance between the mammalian skull and that of the bird, amphibian or fish, in that discriminant characters persist unchanged through transformation.[36] One evidently encounters a graded continuity in form among the various orders of vertebrate animals, as would be expected if evolution occurs along definite lines and within natural constraints.

---

34. Thompson, *Growth*, 318–21.
35. Dewar, Douglas, *The Transformist Illusion* (Hillsdale, NY: Sophia Perennis, 2005), 235–36.
36. Thompson, *Growth*, 321.

Three-Dimensional Co-Ordinate Systems

That the Cartesian system outlined above is not limited to plane geometry is demonstrated by the fact that such a system of co-ordinates can without much difficulty be transferred from two-dimensional figures to three-dimensional bodies, especially in the case of fishes. Thus, when comparing a common "round" fish such as haddock with a common "flat" fish such as plaice, aside from differences in the position of eyes or the number of fins, the chief factor is the broadening out of the plaice's body in a dorsal-ventral direction, as well as its thinning out in another direction. We also find that the high, expanded body of the boar-fish (genus *Antigonia*) or the sunfish (*Mola mola*) is simultaneously compressed vertically; conversely, the skate is expanded from side to side compared to the related shark or dogfish, but simultaneously depressed in its vertical section. This indicates that among fishes the dimensions of depth and breadth tend to vary inversely. Thompson also notes a relation of magnitude between the twin factors of expansion and compression: in the general process of deformation, the volume and area of a cross-section are less affected than its two linear dimensions. Thus with different-looking fishes such as haddock and plaice, they have approximately the same volume when they are equal in length. That is to say, the extent to which the plaice has broadened is compensated for by the extent to which it also became flattened—thus representing an extreme case of the compensation of parts.[37]

It should be kept in mind that fishes, like birds, are subject to strict limitations of form. In a late-nineteenth century study of a number of fishes and whales, which, though mammals, are fully aquatic and thus subject to the same forces as are fishes, it was found that if the areas of their cross-section are plotted against their distances from the front end of the body, the results are similar. The same study also found that the position of greatest cross-section is fixed for all the species observed, namely at a distance of 36% of total body length behind the snout. These rules naturally do not apply to extreme cases such as the eel or the balloon-fish (genus

---

37. Ibid., 323–24.

Diodon), which have materially modified ways of propulsion and locomotion. This evidence implies that hydrodynamical conditions limit form and structure among aquatic vertebrates.[38] In these examples of physical conditions limiting transformations of organic forms, we once again see that in nature, variations are constrained.

Relevance

D'Arcy Wentworth Thompson has become widely recognized as the first bio-mathematician, due to his comprehensive integration of the organic and mathematical realms. The relevance of mathematics in an investigation of the organic realm, and by implication the grounding of the latter in natural law, has been stated with characteristic elegance by the Scottish polymath:

> Every natural phenomenon, however simple, is really composite, and every visible action and effect is a summation of countless subordinate actions. Here mathematics shows her peculiar power, to combine and to generalize.... Growth and Form are throughout of this composite nature; therefore the laws of mathematics are bound to underlie them, and her methods to be peculiarly fitted to interpret them.[39]

According to Thompson, number, order, and position are the threefold clue to exact knowledge in the study of material things. He writes, "For the harmony of the world is made manifest in Form and Number, and the heart and soul and all the poetry of Natural Philosophy are embodied in the concept of mathematical beauty." Not only the movements of the heavenly bodies, but whatever can be expressed by number and defined by natural law must be studied by means of mathematics. "So the living and the dead, things animate and inanimate, we dwellers in the world and this world in which we dwell are bound alike by physical and mathematical law."[40]

---

38. Ibid., 324–25.
39. Ibid., 270.
40. Ibid., 326–27.

Form and Transformation

The Limits of Transformation

In *Origin of Species*, Charles Darwin ascribed such great power to natural selection acting upon variations that he saw no reason why over millions of years terrestrial bears could not be transformed into aquatic whales. He asks rhetorically, "What limit can there be to this power, acting during long ages and rigidly scrutinizing the whole constitution, structure, and habits of each creature—favoring the good and rejecting the bad?" and answers, "I can see no limit to this power, in slowly and beautifully adapting each form to the most complex relations of life."[41] He admitted, however, that if living species were continuously and gradually being transformed into new species, nature should have abounded with transitional forms. Instead, we find a glaring absence of intermediate forms both among living organisms and in the fossil record, but Darwin found a way to account for this absence: "Geology assuredly does not reveal any such finely graduated organic chain; and this, perhaps, is the most obvious and gravest objection which can be urged against my theory. The explanation lies, as I believe, in the extreme imperfection of the geological record."[42] We will now consider the validity of this explanation.

Douglas Dewar was emphatic that Darwin's insistence on the imperfection of the fossil record is at variance with the facts. Already, at the time of Dewar's writing in the middle of the twentieth century, the following were the percentages of discovered fossils representing the mammalian genera alive at the time: land mammals, 61.1% of 408 genera; marine mammals, 75.61% of 41 genera; flying mammals (i.e., bats), 25.56% of 215 genera; for a total 50.14% fossils of 664 living genera. The relatively low percentage for bat fossils was attributed by the British-Indian ornithologist to two factors: flying animals are less likely than others to die from accidents that would result in fossilization of their bodies, and the scarcity of geological exploration in the Tropics, where the large majority of bat genera dwell. If we look at how these findings are distributed on

---

41. Darwin, Charles, *The Origin of Species* (Ware, Hertfordshire: Wordsworth, 1998), 142, 353.
42. Ibid., 132, 213.

the Earth, we see that the percentage of fossils to the living genera of a continent varies from 100% in Europe to 56% in Australia, reflecting the extent of geological exploration of the areas. Dewar concludes that these numbers "indicate that in the course of their existence every genus of land mammal having hard parts is likely to leave its fossil record in the rocks."[43] It was similarly asserted by the Swedish botanist Nils Heribert-Nilsson that by the middle of the twentieth century the fossil record was already so complete that the lack of transitional series could not be explained by the scarcity of material.[44]

By the early twenty-first century, around 150 years after the publication of Darwin's *magnum opus*, vast quantities of fossils had been unearthed, not only in Europe but also in the previously less-explored continents of Asia, Africa and Australia. However, as David Swift remarks, the "missing links" Darwin had hoped for are still missing, since predominantly more members of the same species, more species of the same genera, and more classes of the same phyla have been discovered. Although the Burgess Shale site in Canada provided new phyla and species that prove an exception to this rule, yet the site yielded no intermediates between the previously known phyla. Swift concludes, "So we have many millions of fossils, classified into hundreds of thousands of species, but the radical divisions between groups of organisms persist: the longed-for links remain as elusive as ever."[45] The Russian biologist Lev Berg had earlier pointed out that the lack of transitional forms in the fossil record apply particularly to the transitions between phyla and classes, for example from fishes to tetrapods, from cartilaginous to higher fishes, and even from reptiles to birds, in spite of their apparent affinities.[46]

The writer Perry Marshall notes that this lack of transitional forms can also be observed at the cellular level as a result of symbio-

---

43. Dewar, *Illusion*, 13–15.

44. Thompson, Richard, *Origins. Higher Dimensions in Science* (Mumbai: The Bhaktivedanta Book Trust, 1984), 50.

45. Swift, *Evolution*, 259.

46. Berg, Leo, *Nomogenesis, or Evolution Determined by Law*, trans. J.N. Rostovtsov (Cambridge, MA and London, England: MIT Press, 1969), 347.

Form and Transformation

genesis, a process by which cells merge and cooperate in order to maximize the survival of their organisms. As a result, "multicellular mergers occurred through a number of stages, but single-celled mergers were consummated in a single step. There's no intermediate form between a protozoan and a blue-green algae becoming a cell that does photosynthesis. Cell mergers overturn the Darwinian doctrine of a thoroughly gradual, continuous transition from one species to another." Marshall adds that "there is no transitional form between algae and lichen, and there is no half-merger of two cells. Nature loves shortcuts."[47] We concur with Marshall that these "shortcuts" involve adaptive mutations and other genetic reprogramming, which will be discussed in a later chapter.

According to the modern evolutionary theory, vertebrates evolved from invertebrate chordates, of which lancelets, or amphioxus, and sea squirts, or tunicates, are living representatives. However, Swift has challenged the positing of amphioxus as a vertebrate ancestor, since these cephalocordates have no equivalent to the vertebrate brain, or the neural crest cells and hard tissues of vertebrates. Since there are no known intermediates between cephalocordates and vertebrates, it is nothing more than conjecture that the latter evolved from the former.[48]

Terrestrial animal life is said to have arisen eventually when amphibians evolved from fish towards the end of the Devonian period, around 380 million years ago (hereafter MYA), reptiles from amphibians during the Carboniferous, around 350 MYA, mammals from reptiles during the Triassic, around 220 MYA, and birds from dinosaurs during the Jurassic, around 140 MYA.[49] Regarding the earliest vertebrates, it is claimed that jawed fishes evolved from jawless fishes, by means of the transformation of the gill arch into jaws, for example. However, as Swift notes, no intermediates between jawless and jawed fishes are known. Furthermore, there are several distinct groups among the early jawed fishes, such as the spiny sharks, of which there is no trace of any ancestor nor any form link-

---

47. Marshall, *Evolution*, 125.
48. Cooke, *Animals*, 518; Swift, *Evolution*, 264.
49. Cooke, *Animals*, 28–29.

ing them with modern cartilaginous fish. The morphological diversity among the early jawed fishes is in fact attributed by most paleontologists to polyphyletic (i.e., multiple) origins. Regarding bony fishes, which first appeared in the Devonian, we find two main groups defined by the structure and control of their fins, namely ray-finned and lobe-finned fishes. Both groups appear fully differentiated and at approximately the same time in the fossil record, without any known common ancestor.[50]

Since it is physiologically and mechanically impossible for a ray-finned fish to be gradually transformed into an amphibian, the proponents of gradual transformism allege that tetrapods evolved from lobe-finned fishes, such as the lungfishes now living in Africa and South America. These differ from ray-finned fishes by possessing muscular fins enabling them to cross dry land, as well as lungs to survive in waters with low oxygen levels.[51] An evolutionary series from lobe-finned fishes to tetrapods has been postulated by the molecular biologist Denis Alexander. This series begins with the *Pandericthys*, of which a fossil dating from the mid-Devonian period was found in Latvia. The *Pandericthys* had a pectoral fin and shoulder girdle intermediate between those of a lobe-finned fish and a tetrapod. In Alexander's schema the early tetrapods are represented by the *Ichthyostega* and *Acanthostega*, of which fossils from the late Devonian period were found in Greenland. These early tetrapods possessed the flat-topped skulls of later tetrapods, as well as front and hind limbs with digits. In 2006 fossils of a previously unknown species, *Tiktaalik*, were discovered in northern Canada, also dating from the late Devonian. These crocodile-sized animals had limb-like fins and an elbow joint, which could have enabled *Tiktaalik* to inhabit the shallow waters and adjacent land of what had been a semi-tropical wetland before continental drift set in. This species has (predictably) been posited as a transitional form between fishes and tetrapods.[52]

50. Swift, *Evolution*, 264–65.
51. Dewar, *Illusion*, 39; Cooke, *Animals*, 468.
52. Alexander, Denis, *Creation or Evolution. Do We Have to Choose?* (Oxford, UK & Grand Rapids, MI: Monarch Books, 2008), 128–29.

Form and Transformation

Against *Tiktaalik* as transitional, Swift counters that the structural differences between lobe-finned fishes and the early amphibians weigh heavily against the possibility of a gradual transformation. First, the bones of the posterior fins of the lobe-finned fishes are not attached to the backbone, whereas in tetrapods the hind limbs are thus connected by way of a pelvis. Second, there is no equivalent in the last known fish predecessor to the polydactyl limb of the first amphibians. The third difference pertains to the overall orientation of the limbs, which in fishes are swept backward but in amphibians are directed forward, for movement in water and on land, respectively. Fourth, the ribs of amphibians, dorsal only, are not homologous to those of the lobe-finned fishes, both dorsal and ventral. Moreover, despite certain structural features shared by all three groups of modern amphibians, anurans, urodeles, and caecilians, there is no known common ancestor, since all of these appear fully differentiated in the Jurassic rocks. This fact suggests an origin for amphibians that is polyphyletic, or from more than one lineage, as is the case with the jawed fishes.[53]

Regarding the transformation from amphibians into reptiles, Swift observes that we are once again confronted with fundamental morphological differences between the two, pertaining especially to their respective eggs, reproductive organs, and breathing apparatuses. Their respective scales provide another non-homology, for those of fishes and early amphibians are bony and arise from the dermis, whereas those of reptiles are non-bony and arise from the epidermis. Other evidence against any transformation from amphibian to reptile, lies in the fact that various new reptile groups appear in the fossil record fully differentiated for terrestrial, aquatic or aerial life. Well-known examples of these groups are the dinosaurs, ichthyosaurs and pterosaurs, respectively. Again, these flying and swimming reptiles show no sign of gradual descent from the earlier land-dwelling reptiles.[54] Concerning the egg and the embryo alone, Dewar notes that a whole range of transformations between amphibian and reptile would be required, including the abandon-

---

53. Swift, *Evolution*, 266, 268, 322.
54. Ibid., 268–70, 322–23.

ment of metamorphosis and the setting up of a water supply for the embryo, the formation of new organs such as the amnion and the allantois, and the development of a tooth for breaking the hard egg shell. However, most of these changes would have been useless or even harmful to the embryo until they were more or less complete. We now know that the development of an amniotic egg occurred first among reptiles, facilitating the widespread colonization of land by vertebrates, since these eggs protect the developing embryo within a shell that is permeable to gases such as oxygen and carbon dioxide, while restricting water loss.[55]

Neo-Darwinian evolutionary theory holds that during the early Permian period some reptiles began returning to the water after their land settlement during the late Devonian period. Examples of this are the fully aquatic reptiles of the genus *Mesosaurus*, of which fossils have been found in southern Africa and eastern South America. Incidentally, *Mesosaurus* serves as evidence for the theory of continental drift, since at around one meter long they were too small to have crossed the Atlantic Ocean between the two regions in which they lived. This postulated return of terrestrial vertebrates to the sea has led Douglas Dewar to satirize the Neo-Darwinian position as follows:

> If the doctrine of transformism be true the ways of animals in the past were indeed strange. Fishes which had laboriously turned their fins into legs and become amphibians lost no time in ridding themselves not only of their newly-acquired legs, but of their limb girdles, lock stock and barrel. Again, no sooner had certain amphibians acquired the power of producing eggs that could be incubated out of water than some of them took to living in water.[56]

When considering the transformation of reptiles into birds, we once more encounter fundamental obstacles in the transition from cold-blooded to warm-blooded organisms, from scales into intricate feathers, and from leaping between trees to the mechanism of winged flight, with the concomitant transformation of fore-limbs

---

55. Dewar, *Illusion*, 219–20; Cooke, *Animals*, 26.
56. https://en.wikipedia.org/wiki/Mesosaurus; Dewar, *Illusion*, 44.

into wings—all of these supposedly taking place gradually over numerous generations.[57] The Jurassic fossil of *Archaeopteryx*, with its fully developed feathers, is one of the earliest known birds. Due to its long bony tail, claws on its digits, and teeth, this bird has been proclaimed a transitional form linking birds with reptiles, and is supposed to have derived from dinosaurs known as theropods. However, Dewar has shown that none of the so-called reptilian features of *Archaeopteryx*, such as the skull, non-pneumatic bones, vertebrae, and pectoral and pelvic girdles, are limited to the order of Reptilia. Even the teeth of *Archaeopteryx* are no evidence of a reptilian link, since all the known birds of the Cretaceous and Jurassic periods had teeth, while the modern reptiles turtles and tortoises lack teeth.[58] In addition, the various skeletal differences in pelvis, digits and fore-limbs between theropods and birds count strongly against an ancestral relationship. A further difficulty is that modern birds possess a single thoracic cavity for the distribution of air around their bodies, thereby facilitating the high metabolism related to flying, whereas theropods, along with modern reptiles such as crocodiles, require a thorax divided into two separate chambers. Finally, the most bird-like theropods only appeared in the late Cretaceous, i.e., around 70 million years after *Archaeopteryx*. Given these numerous objections, it is only reasonable to conclude that while *Archaeopteryx* may be intermediate between reptiles and birds in terms of morphology, it is not so according to phylogeny.[59]

Regarding the transformation of reptiles into mammals, we are again confronted with physiological obstacles. For instance, as Dewar argues, the imagined conversion of reptilian jaws and ear bones into those of a mammal disregards the fundamental structural differences between their skeletons. The hypothesis of gradual transformation also fails to explain how or why the Organ of Corti developed in the mammalian middle ear at the same time the jaws and ear bones were being drastically modified. This auditory apparatus, although comprising around 3000 arches forming a tunnel

---

57. Dewar, *Illusion*, 220–23.
58. Swift, *Evolution*, 271, 309; Dewar, *Illusion*, 51–52.
59. Swift, *Evolution*, 271–72.

and supplied by hundreds of nerves, would have been useless in the struggle for survival until it was complete or nearly so.[60] Moreover, similar or even greater changes would have to occur in the remainder of the skeletal structure, as well as in the circulatory, respiratory and digestive systems, the body covering, the transition from cold-blooded to warm-blooded organisms, and the formation of mammary glands, in order to transform a reptile into a mammal.[61] More recently, paleontologists such as Erik Jarvik and Hans Bjerring have likewise contested the presumed transformation of reptilian jaw bones into mammalian ear bones, and even Niles Eldredge of punctuated equilibrium fame admits that the fossil record does not support the transformation of jaw bones into middle ear bones.[62] The geneticist Otto Schindewolf writes that although it is feasible that in the reptilian lineages that led to mammals a gradual reduction of the bones of the lower jaw could have taken place, "the fundamentally decisive, final step—the complete disappearance of these bones or their transformation into elements of the auditory area—must have taken place discontinuously, suddenly, between one individual and the next, during an embryonic developmental stage."[63] The evidence thus suggests that any evolution of reptiles into mammals is physically impossible through gradual transformation, instead requiring a macro-evolutionary mutational "jump."

As is the case with other vertebrate classes, the three divisions of modern mammals, monotremes, marsupials and placentals, appear in the fossil record fully differentiated, with no trace of intermediates. In the case of placental mammals this *ab initio* differentiation pertains to most of its 30 orders. Moreover, those mammals specialized for aerial and aquatic life, namely bats and cetaceans, appear in the fossil record fully adapted. Bats first appear in the Eocene epoch around 50 million years ago, fully specialized for flying and appar-

---

60. Thus providing another example of integrative and irreducible complexity (of which more later).

61. Dewar, *Illusion*, 54–55, 223–25.

62. Swift, *Evolution*, 274.

63. Quoted in Davison, John, *An Evolutionary Manifesto. A New Hypothesis for Organic Change* (2000), 33. http://www.uvm.edu/~jdavison/davison-manifesto.html#order.

ently already using echolocation. There is no paleontological or morphological evidence to indicate that bats evolved out of suggested intermediates such as flying lemurs or flying squirrels, both of which are actually gliders unable to fly. In their turn, whales appear suddenly in the early Tertiary period, fully adapted for swimming with fore-limbs modified as flippers, and with fundamental changes to the ear to enable echolocation.[64]

Since the 1860s the evolution of the horse from small Eocene forms to the modern genus *Equus* has been one of the mainstays of the modern evolutionary theory. The developmental series is postulated to have commenced with *Hyracotherium* (formerly *Eohippus*), around 55 million years ago. During the remainder of the Eocene and then the Oligocene, the Miocene, the Pliocene and the Pleistocene epochs, this fox-sized mammal supposedly developed through a range of intermediate types to the modern horse *Equus*. This series of morphological changes entailed three main components: an overall increase in size, the reduction of the toes into a single large hoof, and the enlargement of the teeth into molars to accommodate the change of feeding habits from browsing to grazing. The fossil record indicates that *Hyracotherium* spread to Eurasia during the Eocene, but became separated from its North American population when the continents split. The Eurasian population became extinct in the Oligocene, but not before it had diversified and also developed characteristics such as increased body size and advanced teeth. This implies that either the same mutations arose in these different populations of *Hyracotherium*, or (more likely) that the diversification arose from segregation of the genes already present in the original population. David Swift concludes that the ancestry of the modern horse can be viewed as a case of limited evolution due to gene segregation and natural selection.[65] In other words, it is conceivable that the living species of the genus *Equus*, for example the horse, donkey, and zebra, developed from earlier forms by means of micro-evolution—that is to say, through a combination of genetic reshuffling and natural selection.

64. Swift, *Evolution*, 274–75; Dewar, *Illusion*, 235.
65. Swift, *Evolution*, 282–84, 292, 377.

According to the modern evolutionary theory, the mammals of the order Cetacea, i.e., whales and dolphins, are descended from terrestrial ancestors that returned to the water around 50 million years ago.[66] Stephen Jay Gould sketched this postulated descent of the cetaceans in an absorbing essay titled "Hooking Leviathan by its Past." The fossil record suggests that around 52 million years ago certain species like modern otters that were fully terrestrial yet at home in the water began to return to the marine habitat of their distant ancestors. Over the next 5 to 10 million years they lost their hind legs, to be replaced by a powerful tail propelling them through water with an undulating dorsal-ventral movement. Their front legs became streamlined flippers for steering in the water, and their ear bones became fused into a stronger jawbone since the mechanism for hearing in water differs from that on land. The fossils of these "transitional" species (*Pakicetus, Ambulocetus, Indocetus* and *Rhodocetus*) have been discovered in India and Pakistan, and fossils of the early whale, *Basilosaurus*, of around 45 million years ago, in Egypt.[67]

Douglas Dewar counters the above hypothesis regarding the development of water mammals, noting that the existence of mammals such as polar bears and sea-otters is problematic, since they are fully at home in water and yet are morphologically similar to their land-dwelling relatives with the exception of the webbed toes of the sea-otter. Since these predators are able to hunt successfully both on land and in the sea, there appears to be no necessity for a complete transformation from a terrestrial to an aquatic mode of life in order to survive in both habitats.[68] It is noteworthy that recent genetic studies have determined whales and dolphins to be ungulates, closely related to the hippopotamus, another land-dwelling form that is yet fully at home in water.[69] As Dewar quips, to assert the return of terrestrial mammals to a fully aquatic mode of life requires belief in two major miracles: the gradual transforma-

---

66. Cooke, *Animals*, 204.
67. Gould, Stephen Jay, *The Richness of Life: The Essential Stephen Jay Gould*, eds. Paul McGarr & Steven Rose (London: Vintage Books, 2007), 615–29.
68. Dewar, *Illusion*, 234.
69. Cooke, *Animals*, 162.

tion of morphological structures over millions of years, and the preservation during this period of countless generations of animals unable to either walk or swim properly. This is in contrast to the theological doctrine of special creation, which requires only one miracle, namely the creation of the first whale. Ironically, the belief in the unlimited power of gradual transformation through natural selection thus requires more belief in the miraculous than is required by belief in special creation.[70]

Ultimately, there is a biochemical argument for the absence of transitional forms in both the fossil record and among living forms. As the biologist John Davison writes, "if specific information was preformed in the evolving genome there would be no need for gradual transformations from one form to another." Moreover, the absence of intermediate forms implies that a primary role for natural selection is to prevent variation and accordingly to maintain the *status quo*.[71] In a later chapter we will present evidence that the role played by natural selection is mostly conservative rather than one of progressive agency, as is asserted in the Darwinian tradition.

---

70. Dewar, *Illusion*, 219.
71. Davison, John, *A Prescribed Evolutionary Hypothesis* (2006), 3. http://www.evcorum.net/DataDropsite/AprescribedEvolutionaryHypothesis.html; Davison, *Manifesto*, 29.

5

The Modern Theory of Evolution

THE TRADITIONAL UNDERSTANDING of evolution as the unfolding of that which has been in-folded has been outlined in an earlier chapter. In contrast to this metaphysically-based concept, modern evolutionary theory or Darwinism in all of its permutations proclaims the transformation of species through contingent events. As defined in the *Oxford Dictionary of Philosophy*, evolution is "the genetic transformations of populations through time, resulting from genetic variation and the subsequent impact of the environment on rates of reproductive success."[1] That is to say, new species are said to arise through the dual mechanism of mutations and natural selection, with the former viewed as random and the latter as preserving those variations conducive to survival and reproduction.

In his recent work, *Evolution 2.0: Breaking the Deadlock between Darwin and Design*, Perry Marshall provides the following useful definitions as they pertain to Neo-Darwinism, the dominant theory of evolution since the mid-twentieth century: random means "occurring without definite aim, reason, or pattern," and mutation means "a sudden departure from the parent type in one or more heritable characteristics, caused by a change in a gene or a chromosome."[2] In addition to these mechanisms of transformation, the modern theory of evolution asserts the gradualism of evolutionary change from simple to more complex forms over immense periods of time. Thus, in the *Oxford Dictionary of Biology*, evolution is defined as "the gradual process by which the present diversity of

---

1. Blackburn, *Philosophy*, 123.
2. Marshall, *Evolution*, 28.

plant and animal life arose from the earliest and most primitive organisms, which is believed to have been continuing for at least the past 3000 million years."[3]

Rather ironically in the light of further developments, the term "evolution" was first applied to Darwin's theory by the sociologist Herbert Spencer and not by Darwin himself. Spencer juxtaposed the development of an individual organism from egg to adult with the transformation of species from protozoon to mammal, describing both as indisputable processes of "evolution."[4] In the 1861 third edition of *Origin of Species*, Darwin appended a brief chapter to credit his predecessors in the development of the theory, such as Jean-Baptiste Lamarck, Robert Chambers, Richard Owen and Alfred Wallace. At that time, Darwin had still not used the term "evolution," although Spencer, whom he viewed as an ally in the struggle against creationism, regularly employed the term. However, with the sixth edition of *Origin of Species* in 1872, Darwin begins to refer to evolution as a "great principle." This tactic was adopted despite the fact that Darwinism is based on the principle of selection, as Etienne Gilson remarks, and not of evolution in its etymological sense. The resulting terminological confusion entails an inevitable metaphysical distortion, as Gilson explains:

> The root of the difficulties is the fundamental indetermination of the notion of evolution. The notion signified something as long as it concerned the development of that which was supposedly enveloped, but Spencer popularized the word in another sense which no one could easily define. Far from being the development of that enveloped, Spencer's system of evolution is a prodigious system of epigenesis where each moment adds something new to the one preceding it.... But whereas one understood an evolution in which the less issued from the greater in which it was contained, that form of evolution in which the greater continually springs from the less is incomprehensible.[5]

This statement by an eminent Catholic philosopher reflects the

3. Quoted in Van Vrekhem, *Evolution*, 11.
4. Bakar, *Criticism*, 159.
5. Gilson, *Aristotle*, 66–69, 77, 103.

metaphysical conviction that the lesser can only proceed from the greater, and not *vice versa*.

As biological component of the modern scientific paradigm, Darwinism rejects any notion of formal and final causality, insisting that matter and mechanism, the material and efficient causes of Hellenic philosophy, are the only factors determining the establishment of biodiversity on Earth. Regarding its rejection of final causality, Darwin's theory of natural selection has been described by D'Arcy Wentworth Thompson as entailing teleology without a *telos*, an adaptation without design, with the "final cause" being little more than the result of sifting the better from the worse—in other words, Darwinian evolution is entirely a mechanistic process.[6] According to Lev Berg, pioneer of the theory of evolution according to natural law (to be discussed in the next chapter), a strictly mechanistic conception of life is only feasible on the assumption that "living machines" can be constructed by inorganic natural forces alone. This was precisely the view of the much-acclaimed physicist Stephen Hawking, who wrote in his 2010 book *The Grand Design* that humans are no more than biological machines and that free will is merely an illusion.[7] Against such full-blown reductionism, Berg aptly counters that "such an assumption is no more justifiable than the expectation of finding in the state of nature a watch, or a steam engine, or a volume of [Tolstoy's] *War and Peace* composed by the blind agency of atoms without any intervention on the part of the human mind."[8]

There is furthermore an ideological underpinning to the modern theory of evolution, one which is mostly ignored in both academic and popular literature on the subject. In his work *At the Edge of History* (1971), the philosopher William Thompson suggests that Darwin was prompted to reason the way he did as a result of living in an empire that placed the white race at the end of a long line of progress that culminated in the Englishman, and with an economic system in which the market was harsh and unrelenting. In his turn, the geneticist Richard Lewontin argues that Darwin took the early nineteenth-

---

6. Thompson, *Growth*, 4.
7. Marshall, *Evolution*, 232.
8. Berg, *Nomogenesis*, 2.

century *political* economy of Adam Smith and his followers and expanded it to include all of the *natural* economy.[9] And the historian Edward Larson attributes Darwin's world-view to the utilitarianism and *laissez-faire* capitalism of his day, he himself having come from a family of successful capitalists. Larson writes, "Natural selection intuitively seemed the right answer to a man immersed in the productive, competitive world of early Victorian England,"[10] which suggests that Darwinism has served from the outset as a scientific ancillary to capitalism.

Darwin and Natural Selection

There exists a popular perception that Darwin arrived at his theory of natural selection purely due to the weight of empirical evidence. However, a rather different picture emerges if we consider what might have been his motives. During his celebrated 1831–1836 voyage on the naval survey vessel, *Beagle*, Darwin lost all faith in the Biblical teaching on God's creation of the world, including the organic realm. The immense biodiversity, including minute adaptations to local conditions, which the English naturalist observed during his circumnavigation of the world, weighed for him against the notion of a divine Creator. Darwin gradually became an agnostic and, by his own confession, remained one until the end of his life.[11]

Darwin published his magnum opus, of which the full title is *On the Origin of Species by Means of Natural Selection or, The Preservation of Favoured Races in the Struggle for Life*, in 1859. It was actually published as an "abstract" of a massive, ongoing work titled *Natural Selection*, prompted by a letter Darwin had received in the previous year from his fellow naturalist Alfred Wallace, who also postulated a theory of natural selection. In the introductory chapter of *Origin*, Darwin outlines his theory as follows:

> As many more individuals of each species are born than can possibly survive; and as, consequently, there is a frequently recurring struggle for existence, it follows that any being, if it vary however

9. Flannery, *Wallace*, 215; Van Vrekhem, *Evolution*, 106.
10. Quoted in Van Vrekhem, *Evolution*, 106.
11. Swift, *Evolution*, 78; Gilson, *Aristotle*, 65; Van Vrekhem, *Evolution*, 21.

slightly in any manner profitable to itself, under the complex and sometimes varying conditions of life, will have a better chance of surviving, and thus be *naturally selected*. From the strong principle of inheritance, any selected variety will tend to propagate its new and modified form.

Darwin added that natural selection constantly scrutinizes every variation, however slight, rejecting the bad ones and preserving the good ones, silently and slowly improving each organic being in relation to its conditions of life.[12]

In his book *Darwin's Dangerous Idea*, the philosopher Daniel Dennett observes that according to Darwin's theory of evolution by means of natural selection, the following conditions are required for evolution to occur: (i) variation, i.e., "a continuing abundance of different elements"; (ii) heredity or replication, i.e., the capacity of these elements to self-replicate; and (iii) differential fitness, i.e., the variability of the copied elements in a given time.[13] Each of these premises, Michael Denton remarks, is practically self-evident.[14]

The paleontologist Henry Osborn claims that the real problem in modern evolutionary theory has always been the origin and development of characters, or individual characteristics of an organism. Strictly speaking, variability refers to characters and not to species of plants and animals as such. He writes, "Since the *Origin of Species* appeared, the terms variation and variability have always referred to single characters; if a species is said to be variable, we mean that a considerable number of the single characters or groups of characters of which it is composed are variable."[15] This caveat should be borne in mind during the remainder of our discussion of Darwin's theory.

Darwin's original theory has been summarized as follows: gradual variation + natural selection + time = evolution.[16] Although Darwin emphasized the role of natural selection in the evolutionary process, he did not claim it as sole cause. In his introduction to *Ori-*

---

12. Darwin, *Origin*, 6, 66.
13. Marshall, *Evolution*, 24.
14. Van Vrekhem, *Evolution*, 36.
15. Quoted in Thompson, *Growth*, 275.
16. Marshall, *Evolution*, 36.

gin of Species (first edition), he wrote that natural selection has been the main, but not exclusive, means of species modification,[17] and expanded this in the 1872 sixth edition, writing,

> Species have been modified ... chiefly through the natural selection of numerous, successive, slight, favorable variations, aided in an important manner by the inherited effects of the use and disuse of parts; and in an unimportant manner, that is in relation to adaptive structures, by the direct action of external conditions, and by variations which seem to us in our ignorance to arise spontaneously. It appears that I formerly underrated the frequency and value of these latter forms of variation, as leading to permanent modifications of structure independently of natural selection.[18]

Here Darwin both recognizes the role of external factors in producing variations, and incorporates Lamarckian elements when he refers to the inheritance of acquired characteristics.

At the beginning of the nineteenth century Jean-Baptiste Lamarck, coiner of the term "biology," suggested that the characteristics acquired by a particular generation are transmitted to their offspring. The French naturalist used the famous argument that if a giraffe was constantly stretching its neck to reach leaves in higher branches, a slightly longer neck would be inherited by its offspring, and in the course of successive generations through incremental modifications a new species would be formed. Lamarck's view was not materialistic, as philosopher and sociologist Steve Fuller remarks, since he conceived life as intelligence striving to transcend its material limits: "Thus, nature comes to be unified through the interaction of living things, each trying to learn from the rest, the legacy of which is then transferred to the next generation."[19]

Like Lamarck, Darwin came to believe that variations arise due to the combined action of the environment and the use or disuse of organs, and viewed these variations as practically unlimited, thereby always providing natural selection with resources. "He believed that

---

17. Darwin, *Origin*, 7.
18. Quoted in Berg, *Nomogenesis*, 16–17.
19. Fuller, *Dissent*, 98.

selection could do everything by small steps over a long period, comparing natural selection with an architect who is forced to build a majestic building with crude stones and yet fulfils the task successfully."[20] In other words, if given sufficient time the interaction between variation and selection always produces new species of animals and plants.

According to Darwin's theory, all species of living beings on Earth have evolved from one or a few common ancestors over immense periods of time. Instead of the prevailing belief at the time that all species of plants and animals had been created separately by Divine edict, nature was now viewed as an organic continuum, in which all animals and plants are descended from very few progenitors. As Darwin wrote, "Therefore I should infer from analogy that probably all the organic beings which have ever lived on this earth have descended from some one primordial form, into which life was first breathed."[21] This view became known as the theory of monophyletic, or single, origin, as opposed to that of polyphyletic, or multiple, origins, as would be postulated by Berg and other scientists.

To be fair, Darwin did acknowledge difficulties regarding his theory, which he discusses in chapters 6 to 8 of *Origin of Species*. These can be summarized as follows: (i) the absence of transitional forms in nature, species instead making their first appearance as well-defined; (ii) the production of complex organs such as the eye by natural selection; (iii) the acquisition and modification of instincts, for example the precise construction of bee cells, through natural selection; and (iv) the sterility of the offspring when species are crossed, compared to their fertility when varieties are crossed.[22] However, this frank admission of problematic aspects did not prevent *Origin of Species* from becoming firmly established as the sacred writ, as it were, of modern evolutionary theory.

Towards the end of *Origin of Species*, Darwin declared that "species are produced and exterminated by slowly acting and still exist-

---

20. Darwin, *Origin*, 102–08; Popov, Igor, "The problem of constraints on variation, from Darwin to the present," in *Ludus Vitalis* XVII: 32 (2009): 202–03.
21. Darwin, *Origin*, 364.
22. Ibid., 132–33.

ing causes, and not by miraculous acts of creation and by catastrophes."[23] With this assertion, Etienne Gilson retorts, Darwin simultaneously confronted Moses, to whom the Genesis creation account was historically attributed, and Georges Cuvier, the famous exponent of catastrophism rather than uniformitarianism in geology.[24] Darwin admitted elsewhere to having two distinct objects in view, namely to prove that species had not been separately created, and that natural selection had been the chief agent of change. Accordingly, in *Origin* he replaced one theology, that of the Divine creation of species, which he detested, with another, evolutionism, which is just another name for the popular cult of natural selection.[25]

It must be admitted that the natural theologians of the Victorian era conceived the Creator in anthropomorphic rather than transcendent terms. The arguments they used failed to distinguish between the First Cause and mediate causes (as we discussed in an earlier chapter), and thus provided easy targets for the more perceptive thinkers among the materialistic and atheistic opponents of natural theology.[26] Among others, the Muslim scholar Osman Bakar notes that the success of Darwin's theory was due mainly to non-scientific factors, such as popular misunderstandings of certain theological doctrines. Bakar adds, "It [Darwinism] became dominant not through its own strength by which it withstood tests, analyses and criticisms but through the weakness of its rivals, those various forms of creationism which were in conflict with each other and which no longer satisfied the positivist's need for causality."[27] However, Darwin's attempt to overcome the deficiencies of natural theology by an approach based entirely on physical principles was equally erroneous, as the mathematician Richard Thompson points out by means of a vivid image:

Speculation resting on a finite set of material observations is

---

23. Ibid., 367.
24. Gilson, *Aristotle*, 176.
25. Van Vrekhem, *Evolution*, 35; Gilson, *Aristotle*, 177.
26. Flannery, *Wallace*, 44.
27. Bakar, *Criticism*, 174.

indeed inadequate to provide knowledge about a supreme transcendental being. But the answer to this problem is not to deny the existence of such a being and to seek explanations solely in familiar physical principles. This is the fallacy of the drunk who lost his keys near the doorsteps of his house but would search for them only under a streetlamp because the light was better there.[28]

Darwin concluded the first edition of *Origin of Species* with a poetic statement:

> There is grandeur in this view of life, with its several powers, having been originally breathed into a few forms or into one; and that, whilst this planet has gone cycling on according to the fixed law of gravity, from so simple a beginning endless forms most beautiful and most wonderful have been, and are being, evolved.[29]

It is pertinent to note that from the second edition onwards, Darwin added the phrase "by the Creator" after the words "originally breathed." This belated reference to a Creator appears to have been a tactical maneuver, since Darwin fervently believed that natural selection does not require any Divine influence to operate efficiently.

The Rise of Genetics

The foundations of what eventually came to be known as the modern evolutionary synthesis were laid step by step during the late nineteenth and early twentieth centuries. Although Darwin had postulated natural selection as the primary mechanism of evolution, he did not present a satisfactory explanation of how new species arise, since he was unaware of the origin of variations. The new science of genetics would strive to fulfil this crucial gap in modern evolutionary theory.

Through his cross-breeding of tens of thousands of pea plants, the Austrian abbot Gregor Mendel (1822–1884) discovered the basic

---

28. Thompson, Richard, *Mechanistic and Nonmechanistic Science. An Investigation into the Nature of Consciousness and Form* (Los Angeles: The Bhaktivedanta Book Trust, 1981), 206–07.

29. Darwin, *Origin*, 369.

laws of biological inheritance. As a devout Augustinian monk, he strove to reveal the original elements and rules of combination used by God to generate the diversity of nature. Steve Fuller adds, "Mendel conceptualized this problem in thoroughly mathematical terms, which laid the foundations for genetics as a statistical science."[30] Yet although he described the laws of heredity in statistical terms, counting the number of hybrid, dominant, or recessive plants among offspring, Mendel was unaware of the mechanism involved. Nonetheless, his discovery that inheritance depended on discrete factors that remained intact during reproduction rather than being blended as had previously been thought, would be of lasting significance for evolutionary theory. Mendel also made a distinction which was to become axiomatic in genetics, namely between the physical traits possessed by a given generation and those it transmits to subsequent generations, or the distinction between phenotype and genotype.[31]

In his turn, the biologist August Weismann (1834–1914) used insights from the new science of cytology, the study of the cell, to formulate his own "germ plasm" theory. This theory holds that all multicellular organisms have two kinds of cells, a small number of reproductive cells that convey information from parents to offspring, and a large number of somatic cells that undertake the ordinary bodily functions. Weismann established experimentally that the reproductive cells, or the germ plasm, cannot be influenced or changed by anything from the rest of the organism or its environment. Consequently, since the hereditary material controls the form of the organism's body but not vice versa, acquired characteristics cannot be inherited, as Lamarck had thought. Weismann held further that variations can only arise through mutation of the germ plasm, since mutations of the somatic tissues are not inheritable. In their turn mutations are said to arise through miscopying of the germ plasm, and therefore the task of natural selection is to prevent the reproduction of mutations that are detrimental to the organism.

---

30. Fuller, *Dissent*, 84–85.
31. Van Vrekhem, *Evolution*, 107; Swift, *Evolution*, 95; Fuller, *Dissent*, 85.

The German biologist thus agreed with the Darwinists on the vital importance of natural selection for biological evolution.[32]

In the late nineteenth century, Mendel's pioneering work on heredity was rediscovered by Hugo de Vries (1848–1935) and other scientists. The Dutch botanist argued that specific traits in organisms are inherited, carried along by particles he called "pangenes," which term would eventually become "genes," indicating specific elements in the hereditary material.[33] Each of these pangenes controls specific characteristics of an organism, and their different combinations explain the varieties displayed by the same species. De Vries conducted a series of experiments with plant hybrids, albeit with the evening primrose rather than peas. He concluded that new species appear suddenly, through discontinuous "jumps" called mutations, and not gradually through slight, successive variations, as Darwin had claimed. Accordingly, these "macro-mutations" were viewed by De Vries, along with his famous contemporary, the geneticist William Bateson, as far more important in the production of new species than the gradual accumulation of minor changes through natural selection.[34] We contend that these two mechanisms underlie macro- and micro-evolution, respectively.

Remarkably, by the beginning of the twentieth century Darwin's theory of evolution had been all but eclipsed by alternative theories on the origins of biodiversity, for example saltationism (the mutational "jumps" of De Vries, Bateson, and others) and orthogenesis, or directed evolution, which we will consider in a later chapter. Then mathematicians such as John Haldane, Ronald Fisher and Sewall Wright came to the rescue of the theory of natural selection by providing it with a mathematical basis, including statistical models. Their work, alongside the experimental work on the fruit fly *Drosophila melanogaster* conducted by Thomas Morgan and his colleagues, gave rise to population genetics. The proponents of the latter differed from Darwin in postulating evolution as taking place

---

32. Van Vrekhem, *Evolution*, 109; Swift, *Evolution*, 91; Marshall, *Evolution*, 292.

33. Although "coding sequence" would be a more accurate rendering of the meaning of "gene"; see Marshall, *Evolution*, 47.

34. Van Vrekhem, *Evolution*, 111–112; Swift, *Evolution*, 92, 95.

The Modern Theory of Evolution

on the level of the species and not of the individual, as he believed. Furthermore, population genetics explains how a favorable trait can spread throughout an entire population by means of gene flow.[35] However, it has more recently been argued that since all evolutionary changes originate in individual chromosomes in individual germinal cells in individual organisms, population genetics has a questionable place in the evolutionary process.[36]

Dobzhansky and the Modern Evolutionary Synthesis

Of lasting relevance for Darwinism was the publication of *Genetics and the Origin of Species* by Theodosius Dobzhansky (1900–1975) in 1937. In this influential work, the Russian-born geneticist and collaborator of Thomas Morgan in his experiments on fruit-flies, combined Darwin's theory of natural selection with the findings of genetic mutation, the latter being the cause of variation. This combination of evolutionary mechanisms would become known as the modern evolutionary synthesis. Over a period of 30 years Dobzhansky simulated the equivalent of 600 years of evolution by exposing thousands of fruit flies, the fast-breeding species *Drosophila melanogaster*, to radiation. Although this produced an immense variety of defects, such as legs growing out of the heads of the hapless insects in the place of antennae, not a single new species or even a new organ was produced. This utter lack of beneficial variations continued to be observed during several further decades of radiation experiments. One of Dobzhansky's famous contemporaries, Richard Goldschmidt, geneticist and founder of the field of evolutionary development, obtained the same lack of constructive results from similar experiments involving moths. Interestingly, when later scientists conducted experiments using much lower levels of radiation, it was found that fruit flies, fungi and bacteria adapted on the cellular level in order to protect themselves against such a threat. As

---

35. Van Vrekhem, *Evolution*, 113–114; Swift, *Evolution*, 98–99; Marshall, *Evolution*, 34.
36. Davison, *Manifesto*, 31. http://www.uvm.edu/~jdavison/davison-manifesto.html#order.

Perry Marshall has observed, these protective adaptations actually challenge the Neo-Darwinian view that evolution is purposeless.[37]

In a later work, *The Biology of Ultimate Concern* of 1967, Dobzhansky divides evolution into three levels or stages: (i) the production of genetic raw materials through mutation; (ii) the formation, through natural selection and Mendelian recombination, of genetic endowments adapted to survive and reproduce; and (iii) the establishment of limits between species through reproductive isolation. However, mutations, i.e., changes in genes and chromosomes, can be either useful or harmful to the organism, and Dobzhansky admits that most mutations are actually harmful, producing defects or diseases, and are sometimes even lethal. Yet a minority of mutations are not harmful but useful to the organism, especially when the living environment of the organism is altered. Through natural selection, harmful genes are then reduced in frequency, while useful genes are perpetuated. Dobzhansky emphasizes that natural selection works not with genes but with organisms, since it is individual organisms that either die without offspring or produce offspring.[38]

In his seminal work *Evolution: The Modern Synthesis* (1942), evolutionary biologist Julian Huxley describes evolution as a series of blind alleys, differing only in their duration. Ironically, given that he was himself a strong proponent of evolution, he appears to challenge the Darwinian view of evolution as an ongoing process. Later scientists such as Pierre Grassé argue that macro-evolution has apparently come to an end, and that only micro-evolution, or variations within a species, continues to occur. The French zoologist also notes that for hundreds of millions of years, no new broad organizational plan has appeared, and that present evolutionary phenomena are no more than slight changes of genotypes within populations. To this the paleontologist Robert Broom adds the observation that since Eocene times, around 50 to 30 million years ago, no new orders of mammals, no new types of birds, and no new families of

---

37. Van Vrekhem, *Evolution*, 116; Marshall, *Evolution*, 30–33.
38. Dobzhansky, Theodosius, *The Biology of Ultimate Concern* (London: Fontana, 1971), 121–22.

plants appear to have evolved. He likewise concludes that all major, i.e., macro-evolution has apparently come to an end.[39]

By the beginning of the twenty-first century the Neo-Darwinian synthesis had come to embrace the following principles: (i) Genes act as units of information in DNA molecules; (ii) the traits of an organism (the phenotype) express the information in its genes (the genotype); (iii) differences in genetic information produce variation in traits; (iv) genetic changes are due to random mutations and are therefore unpredictable; (v) variations in traits in a population occur due to mutational events, a process known as genetic drift; and (vi) natural selection operates on a pool of genetic variants, bringing about survival of greater reproductive fitness.[40] However, the lack of constructive mutations presents a serious challenge to Neo-Darwinism. In spite of extensive laboratory work to generate mutations, not a single case is known where it has led to constructive morphological change. The reason for this absence of beneficial mutations is the improbability of biological macro-molecules such as protein and RNA arising from chance processes. Instead, proteins and RNA were fully functional at the time of their first appearance, and the variations in their amino acid sequences that have occurred since are not susceptible to the selective sorting of natural selection.[41]

Despite the dogmatic insistence by many evolutionists that acceptance of evolution and belief in Divine creation are incompatible, Dobzhansky declared that he was a proponent of both, viewing evolution as God's method of creation. The eminent biologist wrote that creation was not a unique event that occurred around 4000 BC as is held by young earth creationists, but is an ongoing process that began around 10 billion years ago. He also admitted that although the reality of evolution has been established beyond reasonable

---

39. Davison, *Manifesto*, 3–4, 7–8.
40. Hewlett, Martinez, "Molecular Biology and Religion," in *The Oxford Handbook of Religion and Science*, eds. Philip Clayton and Zachary Simpson (Oxford: Oxford University Press, 2008), 176.
41. Swift, *Evolution*, 374.

doubt, its mechanisms require ongoing study and clarification.[42] It is noteworthy that Dobzhansky did not accept the complete rejection of teleology that is common in biological circles, as is evident from the following passage: "Seen in retrospect, evolution as a whole doubtless had a general direction, from simple to complex, from dependence on to a relative independence of the environment, to greater and greater autonomy of individuals, greater and greater development of sense organs and nervous systems conveying and processing information about the state of the organism's surroundings, and finally greater and greater consciousness. You can call this direction progress or by some other name."[43]

Evidently, according to the Neo-Darwinian theory of evolution, "the deterministic mechanisms of natural selection provide systematic guidance for the development of species," thereby counter-balancing the randomness of variations.[44] Since Neo-Darwinism merely proclaims but does not prove that order can be derived from randomness by means of natural selection, it has recently been described by the philosopher Thomas Nagel as "a heroic triumph of ideological theory over common sense."[45] Ultimately, the Neo-Darwinian synthesis fails as a valid theory of evolution because of its emphasis on random mutations, given that most mutations, barring those obtained artificially through radiation exposure, are not random but goal-directed, as we will show in a later chapter.

Gould and Punctuated Equilibrium

One of the most prominent scientific thinkers of the second half of the twentieth century was the American paleontologist Stephen Jay Gould (1941–2002), who together with Niles Eldredge postulated the theory of punctuated equilibrium in a 1972 paper. Through their field research they became convinced that evolution entails long periods of stasis with virtually no change in species, interrupted by

42. Dobzhansky, Theodosius, "Nothing in biology makes sense except in the light of evolution," in *The American Biology Teacher* 35 (1973): 125–29. http://www.pbs.org/wgbh/evolution/library/10/2/text_pop/l_102_01.html.
43. Quoted in Van Vrekhem, *Evolution*, 168–69.
44. Thompson, *Science*, 132.
45. Quoted in Marshall, *Evolution*, 247.

The Modern Theory of Evolution

short periods of explosive evolution in which new species burst into the fossil record. The most celebrated evolutionary explosion is the sudden appearance of nearly all major types of animal life in the early Cambrian seas. Likewise, most of the modern flowering plants appeared within a few million years in the middle Cretaceous period. The paleontological evidence for this has been confirmed by new research on invertebrate and avian paleontology published in 1995, which shows that at least in some lineages the evolutionary pattern was one of millions of years of stasis interrupted by periods of no more than 100,000 years of rapid change.[46]

According to Gould and Eldredge, the fossil record does not agree with Darwin's thesis of gradual transformation, but rather indicates long periods of stasis (from the Greek, "standing still") followed by the sudden appearance of new forms. These new forms appear when small segments of a population are isolated at the geographic periphery of the bulk of a species. Under environmental pressure, favorable genetic variations spread quickly, until the group is established as a new species.[47] A striking example of stasis in the animal world is the coelacanth, a lobe-finned fish of the genus *Latimeria*, of which the ancestors first appear in the fossil record around 400 million years ago. This fish was thought to have become extinct together with dinosaurs around 65 million years ago, until a living specimen was caught off the east coast of South Africa in 1938. Contrary to Darwin's dictum that species are constantly changing due to the endless sifting of natural selection, the coelacanth has remained much the same for several hundred million years. The theory of punctuated equilibrium seemingly provides a plausible explanation for the lack of intermediates between species, while recognizing the principles of population genetics as established in the early twentieth century.[48]

Gould contended that the fossil record is in fact a faithful rendering of what evolutionary theory predicts in the light of punctuated equilibrium. The oldest rocks to retain fossils, these being of

---

46. Denton, *Destiny*, 297–98, 435.
47. Gould, *Life*, 263–66.
48. Van Vrekhem, *Evolution*, 164; Swift, *Evolution*, 288.

prokaryotic cells such as bacteria and stromatolites, are dated to around 3.5 billion years ago. This great age implies that life remained exclusively unicellular for five sixths of its history on Earth, since multicellular organisms only began appearing around 600 million years ago. It was during this time that the vital transition from simple prokaryotic cells to complex eukaryotic cells containing nuclei and mitochondria took place. (As an aside, we should note that recent studies of symbiogenesis have shown that prokaryotes also form symbiotic relationships, thereby blurring the conventional distinction between prokaryotic and eukaryotic cells.)[49] Gould notes that all the major stages in the organization of multicellular architecture for animal life occurred between 600 and 530 million years ago. This period was followed by the remarkable Cambrian explosion starting around 530 million years ago, during which in the space of a mere five million years all but one modern phylum of animal life appeared in the fossil record. From this, Gould conclude that the past 500 million years of animal life amounts to little more than variations on anatomical themes established during the Cambrian explosion.[50]

Throughout his career Gould insisted that there is no direction in evolution, employing the evidence from the Burgess Shale fossils to buttress this stance in his book *Wonderful Life* (1989). In his view, evolution should not be pictured as a branching tree with humans situated right at the top, but rather as a bush with humans being one twig among many, none being more important than any of the others. Gould pointed out the continued existence of the myriads of bacteria and several thousand of beetle species as confirmation of his thesis.[51] Interestingly, while Gould admits that the simplest kind of cellular life arises as a predictable result of the organic chemistry and physics of self-organizing systems, he denies any predictable directions for the later development of life. In addition, by rejecting any notion of direction in evolution, Gould also came to oppose the

---

49. Marshall, *Evolution*, 124.
50. Gould, *Life*, 215–16.
51. Ward, Keith, *The Big Questions in Science and Religion* (West Conshohocken, PA: Templeton Foundation Press, 2008), 65–66.

mythology of progress which has always existed alongside Darwinism. He consequently viewed progress as a delusion arising from the human refusal to accept our insignificance in the face of the immensity of time.[52]

Darwin's insistence on gradualism was based on the geological concept of uniformitarianism (in terms of which every event in the past had to be explained by causes still operating in the present), which Gould had rejected in a 1965 essay, "Is uniformitarianism necessary?" as both incorrect and obsolete. Despite his rejection of gradualism, Gould continued to view himself as a Darwinist until the end of his life. However, his British sparring partner Richard Dawkins claims that Gould's theory is incompatible with Darwinism. With characteristic vehemence Dawkins presents the case in *Unweaving the Rainbow* (2000):

> The extreme Gouldian view ... is radically different from and utterly incompatible with the standard neo-Darwinian model. It also ... has implications which, once they are spelled out, anybody can see as absurd. ... For a new body plan—a new phylum—to spring into existence, what actually has to happen on the ground is that a child is born which suddenly, out of the blue, is as different from its parents as a snail is from an earthworm.[53]

Ironically, Dawkins, for all of his anti-teleological tirades, has not hesitated to employ design-based language in his arguments, such as "selfish gene" and "blind watchmaker." It appears that, for Dawkins, "adaptation" is a secular synonym for "design" and "natural selection" a secular synonym for "God," as the philosopher-sociologist Steve Fuller has wittily remarked.[54]

Although rejecting gradualism, Gould fully subscribed to the contingency on which Darwinism is based. In his 1994 essay "The Evolution of Life on the Earth," Gould discussed the pervasiveness of contingency in the unfolding of life on our planet. He argues that if our particular vertebrate lineage had not been among the few survi-

---

52. Denton, *Destiny*, 296; Van Vrekhem, *Evolution*, 168.
53. Quoted in Marshall, *Evolution*, 133.
54. Fuller, *Dissent*, 157.

vors of the Cambrian explosion that began around 530 million years ago, then vertebrates might not have come to dominate the Earth. Further, if a small group of lobe-finned fishes had not evolved fin bones with a strong central axis capable of bearing weight on land, then vertebrates might never have become terrestrial. Then, of particular importance for mammals, if a large extraterrestrial body had not struck the Earth some 65 million years ago, then dinosaurs might still have been dominant. And finally, if a small lineage of primates had not evolved upright posture on the African savannah between four and two million years ago, then the human ancestry might have ended like chimpanzees or gorillas today.[55] As Gould wrote elsewhere, "We are here because an odd group of fishes had a peculiar fin anatomy that could transform into legs for terrestrial creatures; because the earth never froze entirely during an ice age; because a small and tenuous species, arising in Africa a quarter of a million years ago, has managed, so far, to survive by hook and by crook. We may yearn for a 'higher answer'—but none exists."[56] Thus Gould's (metaphysical) assumption of contingency leads inevitably to the (existential) conclusion of meaninglessness.

Gould uses the memorable yet erroneous argument that if the tape of evolutionary history could be replayed, then the outcome would be radically different. There would be nothing like human beings on Earth, and any other possible biospheres in the universe would have to be different from all the others. However, Gould's notion of radical contingency has been challenged by his one-time collaborator, British paleontologist Simon Conway Morris, who argued that the ubiquity of evolutionary convergence confers at least a degree of predictability as to its outcome: "Convergence simply tells us that the evolution of various biological properties is certainly highly probable, and in many cases highly predictable."[57] Referring to Gould's argument that the history of life is little more than a contingent muddle, in which the mass extinction of one

---

55. Gould, *Life*, 211.
56. Quoted in Van Vrekhem, *Evolution*, 167–68.
57. Conway Morris, Simon, *Life's Solution. Inevitable Humans in a Lonely Universe* (Cambridge: Cambridge University Press, 2003), 223.

group facilitates the survival of another group through sheer luck, Conway Morris writes: "Yet, what we know of evolution suggests the exact reverse: convergence is ubiquitous and the constraints of life make the emergence of the various biological properties very probable, if not inevitable."[58] That is to say, the evolutionary process is guided by constraints and not driven by contingency, accident or luck.

Following Dobzhansky, Gould argued that natural selection occurs on the level of the organism as a whole, i.e., on the phenotype, and not on the genetic level, i.e., on the genotype, as Dawkins claims. Dawkins' reduction of selection to genes struggling for reproductive success has been lambasted by the American paleontologist as both logically flawed and a foolish caricature of Darwin's intention.[59] Gould charged Dawkins and other reductionists such as John Maynard Smith and Daniel Dennett with being "Darwinian fundamentalists," since they failed to appreciate Darwin's insistence that natural selection is not the only mode of evolutionary change. In a review of Dennett's book *Darwin's Dangerous Idea*, Gould takes the American philosopher to task for continuing Dawkins' reduction of Darwin's theory to natural selection alone. Dennett holds evolution to be an algorithmic process, and he accordingly views each of the wonderful adaptations in the organic realm as nothing more than an algorithmic result of natural selection. Biology is explicitly equated with engineering, leading Gould to retort, "If history, as often noted, replays grandeurs as farces, and if T.H. Huxley truly acted as 'Darwin's bulldog,' then it is hard to resist thinking of Dennett, in this book, as 'Dawkins' lapdog.'"[60] Regarding the scope of natural selection, Gould asserts that its operation can explain certain phenomena, such as the behavior of an ant colony, which are above the level of the individual organism. However, it is impossible to attribute all of organic diversity, embryological architecture, and genetic structure to natural selection alone. Also, on the genetic

---

58. Ibid., 283–84.
59. Gould, Stephen Jay, "Darwinian Fundamentalism," in *The New York Review of Books* 44:10 (1997): 3.
60. Ibid., 6, 8.

level natural selection fails to explain why many evolutionary transitions from one nucleotide to another are non-adaptive.[61]

David Swift raises a biochemical objection to the theory of punctuated equilibrium, namely that genes already have to be available in order to operate in small, isolated populations. Since new and useful mutations are exceedingly rare, they have a better chance of arising in a large population. This requirement makes the occurrence of new genetic material in small populations practically impossible. And since small populations do not have enough genetic variability, they can only be a source of new variations, not of speciation. Darwin had made the same argument in *Origin of Species*, writing that although isolation plays a notable role in natural selection, a small isolated population will actually retard the production of new species, by decreasing the chance of the appearance of favorable variations. Swift contends that at best punctuated equilibrium can explain the gaps between related species, but not the transitions between higher taxa such as phyla, classes or orders, all of which appear abruptly in the fossil record.[62]

Criticism

It is a biographical fact that Darwin's theory of natural selection was erected upon the twin pillars of Charles Lyell's geological theory and Thomas Malthus' socio-economic theory. Although Darwin incorporated some aspects of biological reality into his theory, many others were left unexplained, including the mechanisms of inheritance and adaptation. Georges van Vrekhem, a Belgian journalist, contends that apart from making use of the results of artificial breeding and the rudimentary fossil record available at the time, Darwin's theory lacks scientific value.[63] It has also been pointed out that "hybridization is the *only* process mentioned in Darwin's *On the Origin of Species* that is directly observed to originate a new species." But experiments that created new species before the advent of Darwin's theory were the result of hybridiza-

---

61. Ibid., 4, 7.
62. Swift, *Evolution*, 289; Darwin, *Origin*, 81.
63. Van Vrekhem, *Evolution*, 39, 41.

tion, while new species created afterwards were products of symbiogenesis.[64] Both of these mechanisms of directed evolution will be discussed in a later chapter.

In his turn, Wolfgang Smith, mathematician, physicist, and metaphysician, asserts that from the outset Darwin's theory was expounded mainly on *a priori* grounds rather than on empirical evidence. However, due to its fervent propagation by Ernst Haeckel and others, natural selection became widely accepted as an adequate cause of the origin of species. Consequently, Smith writes, Darwinism ceased to be a tentative scientific theory and became a philosophy, if not a religion. Or, as the former Director of Research at the French National Center for Scientific Research, Louis Bounoure, says, the Darwinian theory is strong as dogma, but weak as scientific theory.[65] It has also been noted by Etienne Gilson that the vitality of Darwinism is due to its nature as a hybrid of philosophical doctrine and scientific law, which makes it virtually indestructible. To this David Berlinski, fellow at the Discovery Institute, adds that Darwinism functions as a secular ideology instead of a scientific theory, and, as such, has an inviolability that combines the dogmatism of religion with the entitlement of science. As a matter of fact, Darwinism was never a scientific project in the true sense of the word, contends historian and sociologist Michael Flannery, but rather comprised naturalistic speculations in the service of philosophical materialism. Thus Darwin's reasoning ran consistently from materialism to evolution, and not the other way round, as he claimed in his *Autobiography*.[66]

In recent years we have seen the rise of so-called ultra-Darwinists, or what we might call Darwinian fundamentalists, who reduce the entire evolutionary mechanism to natural selection alone. We thus find the atheist biologists Richard Dawkins and Jerry Coyne marveling at the limitless power of natural selection, while ignoring or downplaying the real mechanisms that generate new characters and ultimately produce new species. In this they follow the example

---

64. Marshall, *Evolution*, 143–44.
65. Smith, *Cosmos*, 82, 84.
66. Gilson, *Aristotle*, 83; Van Vrekhem, *Evolution*, 120; Flannery, *Wallace*, iii, 211.

of August Weismann, a pioneer of genetics, as we have seen, who argues that since acquired characteristics cannot be inherited, it follows that natural selection must be the only major process driving biological evolution.[67] Ultra-Darwinism has also been criticized by Simon Conway Morris for its excessive self-assurance, its distortion of metaphysics to suit ideological purposes, and its quasi-religious fervor.[68] We will now survey some of the scientific and philosophical objections that have been raised by a variety of thinkers against the Darwinian theory of evolution.

Variation *versus* Speciation

One of the most problematic aspects of Darwin's theory is his insistence that variability is practically unlimited and therefore almost inevitably leads in the long run to the appearance of new species. After all, as noted by Igor Popov (writing 150 years after the publication of *Origin of Species*), in spite of all the efforts by geneticists there are no blue-eyed fruit flies, no blue roses, no lupines without alkaloids, no viviparous birds or turtles, and no hexapod mammals.[69] In an early scientific criticism of Darwinism published in 1885, the Russian naturalist Nikolai Danilevsky had already taken the English biologist to task for basing his notion of infinite variability on domesticated birds. Darwin had argued that chickens are highly variable, since humans select them for many different purposes, for example, to be used for meat or eggs, or as gamecocks, or to be decorative. Pigeons are also highly variable, as they have been selected for decorative purposes, while geese are not variable because they are bred only to serve as food. Danilevsky counters Darwin's argument, writing, "Darwin confuses the cause and consequence: the goose has remained constant not because he did not fall in the list of decorative birds, which are appreciated for the beauty and strangeness of form and plumage, but on the contrary, it did not become a decorative bird like pigeons and hens, because it was

---

67. Marshall, *Evolution*, 245, 292.
68. Conway Morris, *Life*, 314–15.
69. Popov, *Constraints*, 211.

The Modern Theory of Evolution

and it is a non-variable species due to its nature!"[70] Danilevsky noted further that despite being decorative, birds such as pheasants and peacocks have remained unchanged, although, had Darwin been correct, they should have displayed a much greater variety.[71]

Astonishingly, despite the title of Darwin's epochal work, it does not offer a single example of speciation. The numerous cases of selection by human breeders, hailed by Darwin as proof of natural selection, produced only varieties and not a single new species. In the words of Titus Burckhardt, "Its [i.e., Darwinism's] advocates present as the start or 'bud' of a new species what is really but a variant within the framework of a determinate specific type."[72] It has to be recognized that artificial selection can significantly alter the phenotype (one only has to think of the morphological extremes among the domestic dog), but none of the products of selection by human breeders have exceeded the species barrier.[73] The French zoologist Pierre Grassé remarked that thousands of years of cross-breeding and selection have produced innumerable variations in animals such as the dog, ox, fowl and sheep, but none of these species has lost its chemical and cytological unity. Moreover, the offspring of dog-breeding are able to interbreed not only with each other but with their wolf ancestor as well. And since the hybrid offspring of such unions are fertile, it confirms that no macro-evolution has taken place, only micro-evolution. The same observation applies to the goldfish varieties bred from the Asiatic carp *Carassius auratus*.[74]

It is also noteworthy that the celebrated breeding experiments with the fruit fly *Drosophila melanogaster* over thousands of generations, involving millions of specimens, and with the mutation rate vastly increased by means of X-rays, failed to produce a single new species. From this, we can conclude that although the reality of micro-evolutionary variations within species cannot be doubted, this does not support the Darwinian claim that new types of plants

70. Quoted in Popov, *Constraints*, 204.
71. Popov, *Constraints*, 205.
72. Van Vrekhem, *Evolution*, 34; Burckhardt, *Cosmology*, 141.
73. Davison, John A., *A Prescribed Evolutionary Hypothesis* (2006): 1. http://www.evcforum.net/DataDropsite/APrescribedEvolutionaryHypothesis.html.
74. Thompson, *Origins*, 46; Davison, *Manifesto*, 9–10.

or animals have arisen through macro-evolutionary transformation. Similarly, the cardiologist Evan Shute argues that to conclude from the observed cases of micro-evolution that "mega-evolution" occurs, is pure conjecture. Instead, micro-evolution demonstrates that there are all kinds of barriers ensuring the stability of classes and orders in the plant and animal kingdoms.[75] We contend that Darwinism fails to account for speciation due to its refusal to distinguish between micro- and macro-evolution. As the geneticist Richard Goldschmidt explains:

> The decisive step in evolution, the first step toward macroevolution, the step from one species to another, requires another evolutionary method than that of sheer accumulation of micromutations.... The fact remains that an unbiased analysis of a huge body of pertinent facts shows that macroevolution is linked to chromosomal repatterning and that the latter is a method of producing new organic reaction systems.[76]

That is to say, a prerequisite for macro-evolution is the production of new genetic material and not only the reshuffling of existing genes, whereas the latter, when producing variations acted upon by natural selection, is sufficient for micro-evolution to occur.

The Origin of Organic Complexity

An early objection to Darwin's theory is that natural selection can provide an explanation for morphological fine-tuning, but not for the appearance of complicated new structures such as the eye or the wing, for which huge mutations would be required. Darwin acknowledged the difficulty of accounting for complex forms in terms of natural selection. However, in *Origin of Species* he then suggests a sequence of gradual changes from a light-sensitive spot to a mammalian eye.[77] To this, Ernst Mayr added that the evolution of the eye ultimately depends on the possession of protoplasm with

---

75. Smith, *Cosmos*, 66–67, 80; Lings, Martin, "Signs of the Times," in *The Sword of Gnosis. Metaphysics, Cosmology, Tradition, Symbolism*, ed. Jacob Needleman (Baltimore, MD: Penguin Books, 1974), 113.

76. Quoted in Davison, *Manifesto*, 13–14.

77. Swift, *Evolution*, 83; Darwin, *Origin*, 143–45.

The Modern Theory of Evolution

photo-sensitivity, since the latter property has selective value from which all else follows by necessity. This reasoning, as Richard Thompson remarks, displays "an abiding faith in the power of natural selection and mutation to effect transmutations in organic form."[78] Furthermore, the postulated evolution of the eye from light-sensitive tissues to the complex vertebrate eye is, according to David Swift, an example of the ignorance concerning the origin of new morphological structures displayed by many biologists. He writes, "They totally ignore the genetic or biochemical implications, and assume that variations can arise and accumulate indefinitely through the imagined plasticity of biological tissues," whereas in reality all morphological changes depend on the biochemistry of the relevant tissues. As a matter of fact, modern biochemistry undermines the Darwinian notion of evolutionary progress through small increments, since for any macro-evolutionary change new genes are required, and not merely the gene shuffling that produces variations, or micro-evolution.[79]

Darwin admitted in *Origin of Species* that his theory would break down completely if any complex organ could be shown to have arisen without numerous, successive, slight modifications. In the light of modern biochemical knowledge, it has to be noted that not only the eye but every organ fails Darwin's test, since each organ depends on complex biological structures in order to function. This also applies to the supposed transition from prokaryotic to eukaryotic cells, because it involves "multiple inter-dependent novelties of molecular biology." Therefore, "every biological macromolecule is, in itself, of a complexity which cannot arise incrementally." Ultimately, although the modern theory of evolution does explain phenomena such as the operation of natural selection at the morphological level, it fails to account for the complexity of molecular biology at the molecular level, in the origin of life and of eukaryotic cells, and in the sudden appearance of new forms in the fossil record.[80]

---

78. Thompson, *Science*, 193.
79. Swift, *Evolution*, 299–300, 315–16.
80. Darwin, *Origin*, 146; Swift, *Evolution*, 386–87.

Sudden Speciation Followed by Stasis

The priority of genetics over selection is evident from the fact that instead of gradual transformation, new species appear suddenly, both in the fossil record and in the experimental laboratory. This phenomenon led the paleontologist Otto Schindewolf to speak of "explosive evolution," since macro-evolution takes place in an explosive manner within a relatively short period of geological time, followed by a slow series of orthogenetic improvements.[81] In the words of the biologist Burton Guttman, "Each speciation event occurs quite rapidly in geological terms, so rapidly that it has sometimes been called 'quantum speciation,' by analogy with the 'quantum jumps' that occur in atoms and molecules."[82] This analogy is not inappropriate, since the organic and inorganic realms are regulated by the same physical and chemical laws, if one leaves aside the role of a transcendent intelligence, at least in the living kingdoms.

The notion of relatively sudden speciation goes against the Darwinian model of gradual transformation over immense periods of time. Darwin had indeed conceded that if it could be shown that numerous, related species had appeared simultaneously, his theory would be invalidated.[83] In order to counter such a rebuttal of Darwin's theory, George Simpson postulated the existence of "fast-rate" evolution alongside "slow-rate" evolution. In his *Tempo and Mode in Evolution* (1944), the American paleontologist suggests that most of the known low-rate organic lines must at an earlier time have been high-rate lines. He employs the example of the bat's wing, which has remained essentially unaltered since the Middle Eocene, apart from some diversification. Simpson concludes, "Extrapolation of this rate in an endeavor to estimate the time of origin from a normal mammalian manus [i.e., hand] might set that date before the origin of the earth."[84] More recently David Swift noted the discrepancy between the fossil record and the molecular clock, i.e., the rates

---

81. Yockey, Francis Parker, *Imperium. The Philosophy of History and Politics* (Torrance, CA: The Noontide Press, 1962), 74; Dewar, *Illusion*, 143.
82. Quoted in Van Vrekhem, *Evolution*, 165.
83. Van Vrekhem, *Evolution*, 162.
84. Quoted in Dewar, *Illusion*, 144–45.

The Modern Theory of Evolution

of nucleotide substitution, as it pertains to mammals and flowering plants, these being the most recently emerged groups of the animal and plant kingdoms. In both cases the molecular data suggest a much earlier origin than indicated by the fossil record. Swift concludes, "There are major discrepancies between the slow evolution that we can reasonably infer from morphologically related species, and the rates that are required to account for large scale evolution"[85]—that is to say, between micro-evolution and macro-evolution.

Interestingly, Louis Bounoure has written that even if finely graduated chains of fossils could be found, it would be impossible to verify any filiation between its links. He adds that it would be erroneous to extrapolate from the facts established by comparative anatomy to posit an actual descent, and that in the whole of the animal kingdom there are no more than five or six series of forms that admit the possibility of organic descent. This suggests that in Ernst Haeckel's celebrated "Tree of Life"[86] made to illustrate Darwin's description of common descent, only the leaves represent taxa of real beings, whereas the large branches and trunk are entirely illusory, contrived to establish a non-existent continuity between groups. As Bounoure observes, these large branches and trunk are only a hypothesis introduced to support another hypothesis (i.e., Darwin's theory), and thus have no more value than a *petitio principii*, or circular reasoning.[87] We suggest that the transformists' conflation of morphological succession with organic descent should be viewed as a biological example of the logical fallacy *post hoc, ergo propter hoc* (after this, therefore because of this).

In his monumental two-volume work *The Decline of the West* (1918–1923), philosopher and historian Oswald Spengler argues that Darwinism has been conclusively refuted by paleontology, since fossils can only be samples representing different stages of evolution, and consequently there ought to be only transitional types and no

---

85. Swift, *Evolution*, 285–86.
86. The concept of which was suggested by Darwin himself; Darwin, *Origin*, 100–01.
87. Smith, *Cosmos*, 67, 71–72.

defined species. Instead, we find stable forms persevering through long ages, forms that "appear suddenly and at once in their definitive shape," without regard for the fitness principle postulated by Darwinism. Spengler added that these stable forms do not appear to evolve towards better adaptation, but become rarer and finally disappear, while quite different forms appear again. In view of these paleontological facts, he suggests a different notion of the evolutionary process: "What unfolds itself, in ever-increasing richness of form, is the great classes and kinds of living beings which *exist aboriginally and exist still, without transition types*, in the grouping of today."[88] This view accords with the metaphysical notion of evolution as the unfolding of inherent possibilities, rather than a transformation determined by external factors.

An American follower of Spengler, the political thinker Francis Parker Yockey, also noted that the fossil record shows only stable forms and no transitional types. The simpler forms, such as bacteria, have not died out or yielded to the principles of Darwinian evolution. Some of them have remained in the same form for hundreds of millions or even billions of years, instead of evolving into something higher.[89] This remarkable phenomenon of living fossils, as Darwin himself termed it, is quite widespread in the plant and animal kingdoms. Some examples are stromatolites among bacteria; cycads, gingko trees, horse-tails and the striking *Welwitschia* of the Namib Desert, among plants; the aardvark, elephant shrews, monotremes and opossums, among mammals; pelicans among birds; crocodiles as well as the tuatara, among reptiles; giant salamanders among amphibians; the hagfish family among jawless fishes; the coelacanth and the mudskipper among bony fishes; many species of shark; and the mantis shrimp, horseshoe crabs and nautilus, among invertebrates.[90] Evidently none of these life-forms have been undergoing transformation into other forms, thereby confirming the morphological stability observed in the fossil record.

---

88. Spengler, *Decline*, 231 (italics in the original).
89. Yockey, *Imperium*, 71.
90. https://en.wikipedia.org/wiki/Living_fossil.

The Analogy of Artificial Breeding

Darwin's observation on the ability of selective breeding to produce new varieties represents the starting point of his theory, as is evident from the opening chapter of *Origin of Species* titled "Variation under domestication." In fairness to Darwin, Wolfgang Smith writes, we should keep in mind that in 1859 such vital factors in the variability of living forms as genes, Mendelian inheritance, mutations, and the endocrine system were unknown, modern biology still being in its infancy. However, it is noteworthy that Alfred Wallace rejected as naively anthropomorphic Darwin's attempt to prove natural selection by means of artificial breeding, remarking that such breeding rather demonstrates *unnatural* selection.[91] In his turn, Lev Berg noted that artificial selection is based on the intelligent will of man, whereas natural selection is based on blind chance. Therefore nature cannot select those individuals which accidentally possess useful variations in order to reproduce. Moreover, the mutations of domestic plants and animals effect complete forms in which the existing genes are regrouped, and not the production of new genes, which is to say that artificial breeding is a matter of micro-evolution, not macro-evolution.[92] Francis Parker Yockey likewise criticized Darwin's analogy between artificial breeding and natural selection, since the products of the former will inevitably be at a disadvantage against their natural varieties. As an illustration of this we need only mention the inability of even large dogs to compete for territory or food against their canine relatives such as wolves or coyotes.[93]

Additional arguments against Darwin's analogy between artificial and natural selection are presented by Douglas Dewar in his work *The Transformist Illusion* of 1955. First, a human breeder can effect changes in animals much faster than can occur under natural conditions, since in nature all but favorable variations are soon eliminated, whereas a human breeder can select any variation regardless of its usefulness to the animal. Second, the variations induced under artificial conditions would most probably not occur in nature; for

---

91. Smith, *Cosmos*, 79; Flannery, *Wallace*, 14.
92. Berg, *Nomogenesis*, 65, 400.
93. Yockey, *Imperium*, 72; https://en.wikipedia.org/wiki/Dog#Competitors.

example, the use of X-rays to accelerate the frequency of mutations. Third, in nature a new variation would most likely disappear through its possessor mating with individuals that do not display the same variation; whereas a human breeder wishing to perpetuate a variation, simply segregates the particular individual for cross-breeding with another individual possessing the same variation. Thus, Dewar concludes, breeding experiments actually afford strong evidence *against* the Darwinian theory.[94]

Also arguing against Darwin's use of the analogy of artificial selection to explain the operation of natural selection, Etienne Gilson points out that in nature new species are formed without a stockbreeder. Darwin attempted to pre-empt this objection by distinguishing between two kinds of artificial selection, namely methodical selection and "unconscious selection."[95] Methodical selection is the well-known practice of stockbreeders and horticulturalists used intentionally to produce new varieties of animals and plants. In contrast, "unconscious selection" is said to occur when stockbreeders spontaneously and randomly choose the most interesting variations to preserve. Gilson counters this, pointing out that such an action remains an exercise of choice and not an accumulation of chances through which nature is supposed to create new species. Thus Darwin's notion of "unconscious selection" reveals itself as completely metaphorical and arbitrary. While artificial selection is not always done scientifically or methodically, it is simply false to claim that it is done unconsciously.[96]

The Improbability of Contingency

Although Darwin himself was not dogmatic regarding the randomness of variations, that the evolutionary process is driven by random genetic mutations became one of the pillars of Neo-Darwinian orthodoxy. These genetic mutations are claimed to be the result of random copying errors. However, at a 1966 symposium held in Philadelphia the mathematicians Murray Eden and Marcel Schützen-

---

94. Dewar, *Illusion*, 154–55.
95. Darwin, *Origin*, 29.
96. Gilson, *Aristotle*, 179–82, 184.

The Modern Theory of Evolution

berger employed probability theory to demonstrate the consistently negative results of random mutations in natural selection. In this context the sociologist Steve Fuller remarks that if natural selection is a chance-based process, then the Darwinian theory of chance is apparently more stringent than that of Aristotle, who viewed chance as two or more independent causal processes of which the outcomes coincide, and thus, as we noted in an earlier chapter, an incidental cause in the attainment of purpose.[97] Since chance is simply a name for the unforeseen meeting of two chains of rigorous causation,[98] chance events do not imply the existence of contingency as is often asserted.

For the sake of fairness, however, we should consider the counter-argument by Theodosius Dobzhansky, namely that the objections to Darwinian contingency are to a large degree based on a misunderstanding. To begin with, not even mutations are random changes, because the mutations in a given gene are determined by the structure of that gene. According to neo-Darwinism, mutations provide the raw material for evolution, for which they are manipulated by natural selection. This latter is only a chance process in the sense that most genotypes or offspring have only relative advantages or disadvantages. Otherwise natural selection is said to be an anti-chance agency, since it makes "adaptive sense out of the relative chaos of the countless combinations of mutant genes," or to word it differently, chance is bridled by the anti-chance agency of natural selection responding to environmental challenges.[99] Nevertheless, although postulating natural selection as an anti-chance mechanism which bestows a degree of rationality on selection, Dobzhansky argues within a purely materialistic framework which is reductionist and thus in our view erroneous.

Earlier, we mentioned the decades-long experiments conducted by Dobzhansky, during which thousands of fruit flies were bombarded with radiation in an effort to accelerate their evolution. Although this resulted in DNA mutations, the experiment failed to

---

97. Marshall, *Evolution*, 28, 36; Flannery, *Wallace*, 43; Fuller, *Dissent*, 169.
98. Ross, *Aristotle*, 80.
99. Dobzhansky, *Biology*, 60, 126.

produce a single new species. Further experiments over the years have established that random mutations actually destroy DNA, at least in the vast majority of cases. It is conceivable, Perry Marshall concedes, that a tiny fraction of useful adaptations could have been caused by random copying errors. Nevertheless, it would still be impossible to prove that such events were random, for the simple reason that there is no mathematical procedure for proving randomness. Thus the random mutation hypothesis, i.e., Neo-Darwinism, cannot be verified, and as such stands in opposition to the scientific model itself.[100]

The improbability of an evolutionary process driven mainly by contingency can also be argued on biochemical grounds. For meaningful genes to arise from random variations of nucleotide sequences, David Swift points out, is in the highest degree unlikely due to the extreme improbability of biological macromolecules arising by chance. He writes:

> There is no way in which such macromolecules can develop their activity progressively, in a trial and error manner comparable with natural selection at the morphological level, because of the minimum number of nucleotides or amino acids that need to be right for the resulting macromolecule to have any utility on which a selective process can operate. This is compounded by the fact that most macromolecules are cooperative—which means that multiple macromolecules must reach some level of utility at more or less the same time and place—which is so improbable a scenario that it must be rejected.[101]

Therefore we should accept that the mutations required for the production of new macromolecules do not occur in a random manner, but are directed towards a specific morphological goal.

A related argument against the Neo-Darwinian insistence on randomness can be based on the fact that the evolutionary process would be impossible without the genetic code found in DNA molecules. As Perry Marshall writes, without code there can be no self-

---

100. Marshall, *Evolution*, 292, 294–95.
101. Swift, *Evolution*, 383.

The Modern Theory of Evolution

replication of cells, without self-replication there can be no reproduction, and without reproduction there can be no evolution. The evolutionary process follows the sequence: genetic code → self-replicating cells → evolution. Furthermore, the genetic code has to obey the basic rules of any communication system in order to function properly. These rules are not derived from the laws of physics, since the latter are invariant, while codes are, by definition, freely chosen. According to Marshall, "Codes are not matter and they're not energy. Codes don't come from matter, nor do they come from energy. Codes are information, and information is in a category by itself." Since the pattern in DNA is a code and all the codes of known origin are designed, this strongly suggests that DNA is designed and not the product of a "happy chemical accident," to which Richard Dawkins attributes the origin of life. Moreover, since communication systems require a reason to communicate in the first place, we have to conclude with Marshall that "codes and life infer not only a designer, but purpose in the universe."[102]

However, the physician Howard van Till suggests that the Neo-Darwinian thesis of random variability could be reconciled with the theological doctrine of Divine creation using insights drawn from chaos theory. He writes: "The dynamically changing state of a physical system traces out a path through 'phase space.' As a consequence of the nature of the system, there are distributed throughout this phase space so-called 'attractors' which lead the system to spend most of its time near these special states. In an analogous manner, it is conceivable that biological species, genera, families, etc., represent genomic attractors in a 'genomic phase space.'" Van Till adds that this concept accommodates both variability (the movement of lineages through genomic phase space) and stasis (the tendency of populations to linger in the vicinity of genomic attractors). He concludes, "In theological terms, the whole array of genomic attractors could be seen as biological potentialities given to the Creation by the Creator, and random genetic variability as the capacity given to DNA to explore that genomic space so that some of those potentialities might be discovered in the course of Cre-

---

102. Marshall, *Evolution*, 178–79, 182, 186–87, 194, 202.

ation's formative history."[103] This notion of genomic phase space and attractors is reminiscent of the theory of punctuated equilibrium, with its recognition of immense periods of stasis, which we associate with micro-evolution, punctuated by brief episodes of rapid speciation, i.e., macro-evolution.

Finally, the Hellenic philosophical notion that Mind/Intelligence and not contingency lies at the origin of the cosmos has been affirmed by a variety of facts drawn from astrophysics, microbiology and computer technology. Based on their researches in these disciplines, the astronomer Fred Hoyle and the mathematician Chandra Wickramasinghe concluded that the complexity of life on Earth could not have been caused by a sequence of random events, but must have been produced by a transcendent cosmic intelligence.[104] And thus is the perennial wisdom of metaphysics confirmed by the provisional findings of physical science.

From Simple to Complex

The Darwinian notion that complex life-forms have evolved out of simple forms has been criticized on various grounds. For instance, Titus Burckhardt writes that the advent of the electron microscope has revealed the astonishing complexity of even unicellular organisms, so that it would be more correct to speak of organic complexity arising from relatively undifferentiated, rather than simple, organisms. The postulated evolution of the simple into the complex has also been opposed by means of information theory. For instance, mathematician Richard Thompson showed that configurations of high information content, such as living organisms, cannot arise with substantial probabilities in models defined by mathematical expressions of low information content, such as the laws of nature in their modern scientific understanding.[105] The implication of this limitation for evolutionary theory is consider-

---

103. Van Till, Howard J., "Basil, Augustine, and the Doctrine of Creation's Functional Integrity," in *Science & Christian Belief* 8:1 (1996): 21–38. http://www.asa3.org/ASA/topics/Evolution/S&CB4-96VanTill.html.
104. Bakar, *Criticism*, 178.
105. Burckhardt, *Cosmology*, 144; Thompson, *Science*, 97.

able, as explained by Osman Bakar: "Complex living organisms, which possess a high information content, could not arise by the action of physical-chemical laws in modern science, since these laws are represented by mathematical models of low information content."[106] Thompson adds that in a physical system governed by simple laws, any information present in the system after transformations took place had to be built into the system from the outset. Since random events are unable to give rise to definite information, Thompson concludes that the existence of a complex order here and now cannot be explained unless we postulate the prior existence of an equivalent complex order or that the information content of the system has been received from an outside source.[107] We consequently have to admit that either the present complexity of the organic realm has been produced by at least an equally complex order, or it has arisen due to external (i.e., transcendent) influence.

Furthermore, the postulated evolution of complex from simple forms has been criticized as violating the law of entropy or the second law of thermodynamics. According to this law, the entropy in a closed thermodynamic system is always increasing—in other words, the level of disorder increases with time. Because the law of entropy is the basis for all physical science, its existence implies that the possibilities latent in evolution are strictly limited, and that evolution has taken place in the face of the laws of nature.[108] Since an increase in entropy entails an increase in the level of disorganization of a system, it is viewed by some as running counter to an evolutionary process characterized by increased complexity and organization. The molecular biologist Denis Alexander argues on the other hand that there are sub-systems effecting a change of one kind of energy into another, such as the photosynthesis of plants and the geothermal activity of vents on the ocean floor, so that it is possible for one system such as organic life to increase in energy while another system such as the sun increases in entropy. However, as Perry Marshall notes, "simply adding energy to a system does not

106. Bakar, *Criticism*, 177.
107. Thompson, *Science*, 98, 134.
108. Blackburn, *Philosophy*, 116; Dewar, *Illusion*, 261.

From Logos to Bios

reverse information entropy in any way, because energy does not create or enhance information." And since information such as the genetic code cannot be derived from matter or energy, it has to be produced by an intelligent being.[109]

The paleontological evidence indicating a more or less ascending but by no means continuous order from apparently simple organisms to ever more complex forms has been explained in metaphysical terms by Titus Burckhardt. On the material plane the relatively undifferentiated always precedes the complex and differentiated, since all matter is like a mirror reflecting the activity of the essences or Platonic Forms by inverting it. This explains why the seed precedes the tree and the leaf bud the flower, whereas in the principial order the Forms pre-exist, beings themselves only following thereafter.[110] Or, as Aristotle says, the order of actual development of things and the order of logical existence are always the inverse of each other (*Parts of Animals*, II. 646).

The Struggle for Existence

Darwin's theory was to a high degree an extension of Malthusianism, taking over its leading idea of a "struggle for existence." Further antecedents of this aspect of Darwinism are mentioned by Francis Yockey, beginning with Arthur Schopenhauer's view of nature as pervaded by a struggle for self-preservation, with the human mind a weapon in the struggle, and sexual love an unconscious mechanism of selection. There is also the theory of descent suggested by the polymath evolutionist Herbert Spencer, who actually coined the slogan "survival of the fittest." Yockey also points out that the religious background of Darwinism was Calvinism, which in an adapted form had been the national religion of England for hundreds of years. In Calvin's teaching the "fit" is called the elect of God, and what Darwin did was to replace election by God with selection by nature. Ultimately, as Lev Berg remarks, the principle of survival of the fittest expresses a self-evident truth, since it means

---

109. Alexander, *Creation*, 138–39; Marshall, *Evolution*, 290.
110. Burckhardt, *Cosmology*, 144.

The Modern Theory of Evolution

the survival of those who survive, i.e., the fittest.[111] The slogan "survival of the fittest" could therefore be viewed as a biological case of circular reasoning, or *petitio principii*.

The Darwinian notion of a struggle for existence was rejected by Friedrich Nietzsche on the grounds that it was more of an assumption than a fact. In *Twilight of the Idols*, the German philosopher admitted that such a struggle does occur, but as an exception. He adds, "The general condition of life is not one of want or famine, but rather of riches, of lavish luxuriance, and even of absurd prodigality—where there is a struggle, it is a struggle for power. We should not confound Malthus with nature."[112] So Nietzsche cautions us to not confuse the undeniable struggle for power among humans with a putative struggle for existence in nature. Yockey concurs that there is in reality no struggle for existence in nature and suggests that this notion is a projection of capitalism onto the animal world. The rule in nature is, according to him, that of abundance: enough plants for the herbivores and enough herbivores for the carnivores. Lord Northbourne also views the notion of a universal struggle for existence as highly anthropomorphic, a projection of the human state of conflict onto the world of nature. The British traditionalist adduces the beauty of flowers in this context: "The floral picture at any rate manifests a joyous superfluity that accords ill with any conception so grim as that of a universal struggle for existence as the influence above all others that made that picture what it is, and has conferred on us the inexplicable and gratuitous benediction of flowers."[113]

Moreover, since cells are capable of effecting their own genetic re-engineering in order to achieve purposeful adaptations (as will be discussed in a later chapter), we might say that cooperation, at least on the cellular level, is more pervasive than conflict in the living kingdoms. As Perry Marshall states so powerfully, "Our bodies are the triumphs of millions of years of stunning genetic innovations,

---

111. Yockey, *Imperium*, 67–68; Berg, *Nomogenesis*, 66.

112. Nietzsche, Friedrich Wilhelm, *Twilight of the Idols*, trans. Antony Ludovici (London: Wordsworth, 2007), 55.

113. Yockey, *Imperium*, 69–70; Northbourne, *Progress*, 92.

mergers, and partnerships. The world is what it is because of ingenious systems and designs. Not randomness. Not luck. Humans destroy the Earth; cells rebuild it. Cells are smarter than humans. Cooperation trumps survival of the fittest."[114]

Utility *versus* Teleology

During the nineteenth century English thinkers such as Jeremy Bentham and John Stuart Mill postulated the doctrine of utilitarianism, which taught that the ultimate purpose of life is the maximization of utility or happiness. This idea of utility came to permeate the scientific disciplines of biology and geology, as we can see in Scottish geologist Charles Lyell's work dealing with the formation of geological strata and in Darwin's on the origin of species. In place of the mechanisms of catastrophe and metamorphosis postulated by some of their predecessors, "they put a methodical evolution over very long periods of time and recognize as causes only scientifically calculable and indeed mechanical-utility causes."[115] Francis Yockey points out that the utilitarian aspect of Darwinism is subjective, since the utility of an organ is relative to the use made of it. Until it reached perfection, a slowly-evolving organ (such as a hand) would actually be disadvantageous to the organism, and the process by natural selection could take countless generations. Yockey points out that even a lack of something can be useful, for a lack of one sense can develop others. Physical weakness can stimulate intellectual development, as has often been observed, and an absence of one organ stimulates other organs to compensate, as for example in endocrinology. Spengler concludes from all this that the assumption of utility or other visible causes for biological phenomena has no support in actuality.[116]

In his summary of the theory of natural selection titled *Darwinism* (1889), Alfred Wallace contends that humans differ from other animals not only in degree, but also in kind. He argues that human

---

114. Marshall, *Evolution*, 276.
115. Spengler, *Decline*, 230–31.
116. Yockey, *Imperium*, 72–73; Spengler, *Decline*, 231–32.

The Modern Theory of Evolution

activities such as mathematics, abstract reasoning, art, and music cannot be explained in terms of natural selection or utility, but instead serve as evidence of a spiritual essence capable of further development under favorable conditions. He also questions whether Darwin's theory can account for the origin of the mind, asking what relation there is between an improvement of the mathematical faculty and the struggle for life, or the survival of one's tribe, nation, or race. He concludes that the mind cannot have been produced by natural selection. Arguing along similar lines against the Darwinian notion of utility, the philosopher of science William Thompson poses the question of why humans evolved a brain far more complex than that needed for survival.[117]

It appears that Darwin was ambivalent concerning teleology. On the one hand, he admired beauty in nature, sensible, as in the colors of birds, as well as intelligible, as in the mutual adaptation of bodily parts one to another. He attributed sensible beauty to utility, notably in the phenomenon of sexual selection. Since the beauty found in adaptations is a means to an end, these adaptations being intelligible only from the point of view of their final result, Darwin here approaches the notion of final causality.[118] However, in his *Autobiography* the English naturalist declared that the argument from design, i.e., teleology, has been replaced by the law of natural selection: "There seems to be no more design in the variability of organic beings and the action of natural selection than in the course in which the wind blows. Everything in nature is the result of fixed laws."[119] And the philosopher John Stuart Mill remarked in an early review of *Origin of Species* that Darwin had literally denied the intelligibility of nature.[120] Evidently, Darwinism has been anti-teleological from the outset, with no allowance made for final causality in the organic realm.

---

117. Flannery, *Wallace*, 20–21, 213, 215.
118. Gilson, *Aristotle*, 97–98.
119. Quoted in Swift, *Evolution*, 81.
120. Fuller, *Dissent*, 88.

Progress and Evolution

Viewed from a traditionalist perspective, Darwinism is a by-product of the modernist myth of progress, which conceives human progress in strictly material terms. According to the Iranian philosopher S.H. Nasr, several factors contributed to the modern Western belief in progress through material evolution. To begin with, in the Renaissance West, the traditional Christian teaching that humans are born in order to become more than human (notably expressed in the Greek Patristic doctrine of *theōis* or becoming like God) was challenged by a view limiting mankind to the purely human and earthly. As Nasr writes, "Renaissance humanism, which is still spoken of in glowing terms in certain quarters, bound man to the earthly level, and in doing so imprisoned his aspirations for perfection by limiting them to this world." Shortly thereafter, territorial conquests in the Americas, Africa and Asia brought great wealth into Western Europe, thus stimulating a belief in material progress. Some forms of Protestant theology even associated economic activity with moral virtue, thus contributing to the rise of capitalism in the Western world.[121] The causal link between Protestantism, especially Calvinism, and capitalism was famously postulated by the German sociologist Max Weber in his work *The Protestant Ethic and the Spirit of Capitalism* (1905). By the late twentieth century the conflation of evangelical Protestantism with material wealth had reached a nadir in the so-called "prosperity gospel," which critics aptly referred to as "Cadillac theology."

The increasing secularization of the West from the Renaissance onward resulted in the replacement of the transcendent by the immanent, the spiritual by the material, and belief in the hereafter by a focus on the here and now. It is also interesting to note that the belief in human progress aroused such fervor that it began serving as a pseudo-religion, and was later interwoven with evolutionism. Concomitantly, as Nasr points out, Christian teaching on the linearity of historical time was transformed into faith in human progress through historical change, as expounded by Hegel and

---

121. Nasr, *Progress*, 45.

The Modern Theory of Evolution

Marx, and the rise of fully-secularized utopianism during the eighteenth and nineteenth centuries led to an almost messianic zeal to establish a perfect social order in this world, whether through revolution or evolution.[122]

Of decisive significance in the development of the progress myth was the so-called scientific revolution of the seventeenth century, during which the symbolical world-view still affirmed by Renaissance thinkers such as Marsilio Ficino, Pico della Mirandola, Nicola Flamel, Leonardo da Vinci, and Giordano Bruno, was replaced by the mechanistic world-view of Johannes Kepler, Galileo Galilei and Isaac Newton, although the latter still recognized that the observed regularities of the planets provided evidence of a designer.[123] The philosophical groundwork for this "revolution" was provided by René Descartes' radical distinction between *res cogitans* or thinking substance and *res extensa* or extended substance, i.e., material things, which resulted in a complete divorce of mind from matter. Thus was created a new science, in which matter was reduced to a mere thing and everything could be explained by means of quantitative laws. It should, however, be noted that matter in this reductionist view has nothing in common with Greek *hylē* or Latin *materia*.[124]

The correlation between evolutionism and the progress myth has also been emphasized by the Sufi scholar Martin Lings. Employing a familiar visual analogy, Lings writes:

> In this context the theories of evolution and progress may be likened to the two cards that are placed leaning one against the other at the "foundation" of a card house. If they did not support each other, both would fall flat, and the whole edifice, that is, the outlook that dominates the modern world, would collapse. The [modern] idea of evolution would have been accepted neither by scientists nor by "laymen" if the nineteenth-century European had not been convinced of progress, while in this century evolution-

---

122. Ibid., 46–48.
123. Ibid., 49; Marshall, *Evolution*, 211.
124. Nasr, *Progress*, 49.

ism has served as a guarantee of progress in the face of all appearances to the contrary.[125]

As a matter of fact, one of the founders of the modern evolutionary synthesis, Julian Huxley, did not limit his view of evolution to biology but expressed it as an all-embracing world-view. In an essay titled *Evolution, Culture and Biology* (1955), Huxley defined evolution as "a self-maintaining, self-transforming, and self-transcending process, directional in time and therefore irreversible, which in its course generates ever fresh novelty, greater variety, more complex organization, higher level of awareness, and increasingly conscious mental activity."[126] This combination of a deficient notion of evolution with a mythical progress also serves to explain the messianic zeal with which the Neo-Darwinists habitually attack opposing viewpoints.

Furthermore, it is no coincidence that modern evolutionary theory arose in the nineteenth century, which marks the nadir of metaphysical tradition in the Western world, as Nasr stated in his 1981 Gifford lectures, published as *Knowledge and the Sacred*. Among the few thinkers who retained at least a connection to the metaphysical point of view were Johann Wolfgang von Goethe, Thomas Taylor, William Blake, and Ralph Waldo Emerson. In the general absence of metaphysical thought in the West, theological notions such as the immutability of species, divine archetypes, creation, and design or purpose in nature, all came to be understood, or rather misunderstood, according to their popular formulations. To give just one instance, the theological doctrine of *creatio ex nihilo* or creation out of nothing was understood literally and as such attacked by the evolutionists. Understood metaphysically, this doctrine should be seen as creative emanation, which explains the real meaning of *ex nihilo*.[127]

The twin notions of progress and Darwinian evolution arose within a Judeo-Christian cultural context, which can be explained

---

125. Lings, *Signs*, 112.
126. Quoted in Dobzhansky, *Biology*, 36.
127. Bakar, *Criticism*, 163–64.

in part because Christianity is the most history-conscious among the world religions. However, because of their belief in an original state of perfection followed by a fall that had catastrophic effects, neither Christianity nor Judaism taught any kind of progressive evolution towards a perfect state. In the nineteenth century the ideas of German philosophers Immanuel Kant, Johann Gottfried von Herder, and Johann Gottlieb Fichte regarding the evolution of human society culminated in the system of Hegel, which viewed history as the progressive manifestation of the Spirit. Marx in turn replaced the role of Spirit by economic forces, especially production. In a striking coincidence, Marx and Darwin postulated their evolutionary theories of society and life, respectively, at approximately the same historical juncture.[128]

Materialism *versus* Levels of Being

According to the materialistic world-view, physical matter is the only reality, and therefore everything—thought, emotion, and will included—can be explained in material terms. This two-dimensional view of reality is closely related to the notion of naturalism, which teaches that the natural world is a closed system. Proponents of naturalism therefore declare that nothing supernatural can exist, and thus God cannot exist. The only reality is that which can be explained by the methods used by the natural sciences.[129] It is on this combined scaffolding of materialism and naturalism that modern evolutionary theory has been constructed. However, we should recognize that methodological naturalism is a valid assumption in certain arenas such as laboratory work, where scientific experiments are testing natural processes, not divine interventions. "As long as we respect this," writes Perry Marshall, "then science and God are not incompatible.... Rather, belief in God reinforces order and rationality in science."[130]

Frithjof Schuon also criticized modern evolutionism for its exclusive horizontality. The "evolutionist error," as he calls it, is rooted in

---

128. Dobzhansky, *Biology*, 37–38.
129. Marshall, *Evolution*, 205; Blackburn, *Philosophy*, 245.
130. Marshall, *Evolution*, 218.

the materialists' belief that only the horizontal and natural exist, while the vertical and supernatural are rejected. Thus, "instead of conceiving that creatures are archetypes 'incarnated' in matter, starting from the Divine Intellect and passing through a subtle or animic plane, they restrict all causality to the material world, deliberately ignoring the flagrant contradictions implied by this conceptual 'planimetry.'"[131]

Moreover, the evolutionist doctrine that the origins of life, sentience, and intelligence can be explained in materialist and mechanistic terms represents a blatant denial of all intelligible reality, as Osman Bakar has correctly stated. In contrast, traditional metaphysics recognizes various degrees of reality, all grounded in the Divine Principle. In an ascending order, reality consists of the material or sense-perceptible level, the subtle or psychic level, the intelligible or supra-formal level, Being as ontological Principle, and, finally, the Absolute, which is Beyond-Being. Proponents of modern evolutionary theory appear to be ignorant concerning these levels of being. Bakar writes that they fail to recognize that "between the organism that simply lives [e.g., plants], the organism that lives and feels [e.g., animals], and the organism that lives, feels and reasons [e.g., humans], there are abrupt transitions corresponding to an ascent in the scale of being and that the agencies of the material world cannot produce transitions of this kind."[132] These three levels of being evoke Aristotle's levels of soul, namely the nutritive, the sensitive and the rational, as we have noted in an earlier chapter. Between these levels of soul, and hence of being, there are ontological discontinuities which cannot be reduced to material causes.

The modern theory of evolution asserts that, under certain conditions, matter produces life and animate matter produces spirit. It is further claimed by proponents of mechanistic evolution that the final conditions under which animate matter produces spirit are located in the functioning of the brain. In his *Postmodern Metaphysics*, philosopher and theologian Christos Yannaras posits a herme-

---

131. Schuon, Frithjof, *Sophia Perennis and the theory of evolution and progress* (2001): 1. http://www.frithof-schuon.com/evolution-engl.htm.

132. Bakar, *Criticism*, 169–70, 179.

neutic challenge to the theory of evolution by underlining the discontinuity from inanimate to animate matter and from brain function to specifically human spiritual activity. He argues further that the claim of random and accidental genesis of animate matter does not leave the modern theory of evolution with logical space.[133] The independence of consciousness from matter is also implied by the mathematician Norbert Wiener, inventor of cybernetics, when he states that there are three basic entities in the universe, namely matter, energy, and information. Whereas matter and energy are interchangeable, as Einstein indicated in his famous equation $E = mc^2$, information is not an emergent property of matter but can only be derived from consciousness. And, according to the theory of biocentrism as expounded by the biologist Robert Lanza, the existence of matter is preceded by the principle of consciousness.[134] There can be no doubt that all the thinkers in the Platonic tradition of the past two thousand years would have agreed.

Not only is information not derivable from matter or energy, it results from freedom of choice, as Perry Marshall notes. This property enables freely chosen decisions that cannot be derived from the laws of physics, and information capacity is precisely such a capacity for choice. Since information is communicated among cells by means of sophisticated cellular language (of which more in a later chapter), this sharing of information enables them to overcome the regressive effects of physical laws such as those governing gravity and entropy. An authentic theory of evolution therefore has to recognize that the unfit adapt, that order and structure increase, and that cells exert control over their environments. Marshall concludes that since materialism is unable to explain the origin and nature of information, or the ability to create a code or language, it cannot explain evolution.[135] After all, the genetic code is primarily information and it can therefore not be reduced to material factors, as is done in Neo-Darwinism.

---

133. Yannaras, Christos, *Postmodern Metaphysics*, trans. Norman Russell (Brookline, MA: Holy Cross Orthodox Press, 2004), 73–77.

134. Marshall, *Evolution*, 203.

135. Ibid., 206–07.

Transformation *versus* Manifestation

In his essay *On Earth as it is in Heaven*, contemporary American Traditionalist scholar James Cutsinger presents a number of reasons for rejecting Darwinian evolutionism. He writes that Darwinian transformism is explicitly limited to the material level, which is the least real dimension of the cosmos and hence the least intelligible. When it attempts to explain the more by the less, the transformist cosmogony implies that the lowest explains the other levels and that ultimately something is derived from nothing. Thus is obtained a materialist and atheist parody of the Christian doctrine of God's creation from nothing, namely the creation of something *from* nothing *by* nothing. Furthermore, by mistaking temporal succession for ontological causation, transformism succumbs to the logical fallacy of *post hoc, ergo propter hoc*, as we have argued above. In the metaphysical understanding, the order in which the various species of plants and animals have been disclosed in matter over time is the reverse of their supra-temporal order as archetypes or Ideas. It is therefore incorrect to state that mammals are derived *from* reptiles, Cutsinger remarks, but they are as it were derived *through* them. In addition, the Darwinian theory is blind to the fundamental distinction between form and shape, or the Hellenic distinction between intelligible *eidos* and sensible *morphē*. Whereas a species is dependent on shape, and shape is a function of surface, form denotes the ontological link between an organism and its intelligible archetype. Therefore, Cutsinger writes, "A truly adequate cosmological explanation of the world is an explanation of forms and their hierarchical order, and of That which they wish to express." Finally, Cutsinger points out that Darwinism fails to account for the reality of mind or *nous*, reducing it to the result of physical, chemical and/or biological operations. In reality, mind transcends all natural processes, since it proceeds from the Source of all reality.[136]

Cutsinger concedes that none of these objections to Darwinism prevents the metaphysical thinker or theologian from recognizing the sequential appearance of life-forms on the Earth, as well as the

---

136. Cutsinger, *Earth*, 5–9.

ongoing variability in the physical constitution of plant and animal species. The empirical evidence provided by the geological record, the techniques of radiometric dating, and the results of artificial breeding in no way repudiates the traditional cosmology and metaphysics. As Schuon remarked, facts are always compatible with principles.[137]

Conclusion

We have surveyed a wide range of objections to the modern theory of evolution. Aware that we must guard against the all-too-human tendency of throwing out the baby with the bath-water, to speak colloquially, we concur with David Swift that Darwin should be credited with recognizing the importance of variations, natural selection, and evolution within limits, that is, micro-evolution. However, while Darwin was correct in his identification of the existence of variations and the role played by natural selection, he was wrong in the unlimited extrapolation thereof, as if these factors could account for all morphological changes. Although micro-evolution is not only possible but inevitable, it can only take place within the limits imposed by the existing genetic material. Such evolution is substantially confined to genetic recombination and selection, and to make the leap to macro-evolution is nothing more than speculation, since the latter requires the formation of new, useful genetic material.[138]

Darwin and his followers therefore err in summarily extrapolating from micro-evolution to macro-evolution, since in their veneration of the power of natural selection, they neglect to consider the origin of variations or the limits to variability. Thus, while Darwinism provides a feasible empirical explanation of micro-evolution, it fails as an explanation of macro-evolution, for which the production of new genetic material is required. We contend that macro-evolution is more feasibly explained as a result of regulated, directed, and convergent evolution, which we will discuss in the next three chapters.

137. Ibid., 10, 22.
138. Swift, *Evolution*, 381–83.

6

Evolution According to Natural Law

AN EARLY EXPONENT of organic evolution as a process directed by natural law was Robert Chambers, in his *Vestiges of the Natural History of Creation* (1844). In this epochal work the Scottish writer argued that to attribute the production of the progenitors of all existing species to personal exertion by God is to anthropomorphize the creative power of the Deity. Chambers argues:

> Some other idea must then be come to with regard to the *mode* in which the Divine Author proceeded in the organic creation.... We have seen powerful evidence, that the construction of this globe and its associates, and inferentially of all the other globes of space, was the result not of any immediate or personal exertion on the part of the Deity, but of natural laws which are expressions of his will.... More than this, the fact of the cosmical arrangements being an effect of natural law, is a powerful argument for the organic arrangements being so likewise, for how can we suppose that the august Being who brought all these countless worlds into form by the simple establishment of a natural principle flowing from his mind, was to interfere personally on every occasion when a new shell-fish or reptile was to be ushered into existence on *one* of these worlds? Surely the idea is too ridiculous to be for a moment entertained.[1]

This is indeed an elegant argument against what would eventually become known as literal creationism.

Instead of such ongoing Divine intervention, Chambers suggests that the properties of the elements at the moment of their creation

---

1. Quoted in Denton, *Destiny*, 269–70.

adapted themselves to an infinity of useful purposes. Since both the cosmos and the world of life are produced by natural laws, Chambers viewed all reality, biological and physical, as one immense interconnected Divine artefact. Moreover, he anticipated an ongoing biological debate by postulating an analogy between embryology (or rather ontogeny, the development of an individual organism from an egg cell) and phylogeny (the development of a species from an original progenitor), since both processes are determined by the laws of nature. Although Chambers was vehemently attacked by members of the Victorian scientific establishment, his work appears more relevant than ever in the light of today's cosmology and molecular biology.[2]

Using Platonic cosmogony as his inspiration, Richard Owen, in his work *On the Nature of Limbs* (1849), states that all the basic recurrent forms of both the organic world, such as the pentadactyl design of vertebrate limbs, the body plans of major phyla, or the forms of leaves, and of the inorganic world, such as atoms and crystals, represent material manifestations of a finite set of immaterial archetypes.[3] Apparently, Owen agreed with Chambers that, just as God assembled the heavenly bodies through natural laws, so also is the organic realm the result of natural laws. An indisputable instance thereof is the law of formation, duration, and dissolution, which in the organic realm is the law of birth, life and death. Lord Northbourne adds that this law "is applicable to all systems, from man to amoeba, and from spiral nebula to atom."[4] It is noteworthy that the account of the inorganic realm that is widely accepted today is one based on law, as we see for example with the atom building rules, the laws of crystallography, and the laws of chemistry.[5] Now, that the same laws apply to the inorganic and organic realms is a necessary corollary of the metaphysical principle that the whole of the sensible world, animate and inanimate, receives its being from the intelligible world in which it participates.

2. Denton, *Destiny*, 270–71.
3. Denton, *Protein*, 325.
4. Northbourne, *Progress*, 30.
5. Denton, *Protein*, 327.

After the 1859 publication of Darwin's *magnum opus*, Platonic cosmogony was more or less discarded. As Michael Denton and his co-authors correctly observe, the machine or artefact became the new model of organic form, contingency replaced necessity, and natural selection replaced natural law.[6] We should note, however, that, as we see from the closing pages of *Origin of Species*, Darwin does recognize natural laws as secondary causes, writing that the whole organic realm was produced by the operation of laws "impressed on matter by the Creator." These laws are: growth with reproduction; inheritance; variability due to external conditions, use and disuse; a ratio of population increase leading to a struggle for life; and ultimately natural selection, which entails both the divergence of character and the extinction of unfit forms.[7] In a recent study of Darwin's use of theology, Stephen Dilley argues that Darwin believed in a deist god who created the universe and then abandoned the world to the operation of fixed natural laws.[8] Evidently, Darwin conceives of natural law as purely mechanical in nature.

Denton notes that it is rather ironic that Darwinists have adopted the same metaphor as that used by their creationist opponents. Both groups view organic order as contingent and artefactual, like the order of a machine created either by God or by a "blind watchmaker," the latter being the title Richard Dawkins uses for one of his books. As an aside, Dawkins' title is a contradiction in terms, since a watch is an example of a complex integrated system that can be changed only by intelligent direction. To return to the artefactual metaphor, this new understanding of organic form inaugurated the era of modern biology and changed the whole explanatory framework of biological science. This change was revolutionary, the entire organic realm coming to be seen as a tiny, finite set of all possible forms, drawn by selection from a potentially infinite set during the evolution of life on Earth.[9]

6. Ibid., 328.
7. Darwin, *Origin*, 368–69.
8. Marshall, *Evolution*, 254.
9. Denton, *Protein*, 329–30.

Evolution According to Natural Law

In contrast, non-Darwinian evolutionary theory asserts that since organic forms arise due to the laws of physics, the whole evolutionary pattern has to be conceived as pre-determined by natural law. Thus, in his *Anatomy of Vertebrates* of 1866, Owen writes that the path of evolution was pre-ordained due to an innate tendency of nomogenously-created (i.e., generated by law) protozoa to rise to higher forms. Moreover, because adaptations are *ad hoc* functional modifications of primal patterns given by natural law and not contrivances designed by God, organisms cannot be expected to exhibit perfect adaptations.[10] With this ingenious argument Owen affirmed both the transcendent origin of life and the observed phenomenon of imperfect adaptations.

The first comprehensive scientific assault upon Darwin's theory of natural selection came from Russian zoologist Lev Berg, whose *Nomogenesis, or Evolution determined by law* was published in 1922. In its 400-odd pages, Berg not only takes Darwin's theory to task, but simultaneously develops an alternative theory of evolution, which he terms nomogenesis (from the Greek *nomos*, law and *genesis*, origin). The introduction to the English translation published four years later was written by D'Arcy Wentworth Thompson, whose monumental work on organic form we touched upon in an earlier chapter. A further English edition of *Nomogenesis* was published in 1969, to which Dobzhansky contributed the foreword. As Dobzhansky explains, "nomogenesis is one of the autogenetic theories of evolution, which postulate that evolution is an unfolding of pre-existing rudiments or potentialities rather than a series of adaptive responses of living species to their environments."[11] The following categories of evidence were adduced by Berg in favor of evolution determined by law: (i) facts from paleontology and comparative anatomy; (ii) the phenomena of convergence, parallelism, and analogous variations; and (iii) the process of individual development, i.e., ontogeny.[12] We will now consider arguments presented by Berg as well as more recent scientists.

10. Ibid., 327–28.
11. Berg, *Nomogenesis*, x.
12. Ibid., 111.

The Origin of Adaptations in Organisms

Berg writes that, "as a general rule, organisms, so far as it lies in their power, respond to stimuli in a purposive manner."[13] By purposive, he means acting so as to prolong the life of the individual organism or the species. Adaptations provide organisms with the ability to accomplish purposive acts, and these presuppose both adequate structure and adequate function, that is, the capacity to utilize a given organ. Exceptions to purposive adaptation such as the flight of moths towards fire, the extermination by the female of her progeny, same-sex attraction, abortive regeneration, and anaphylaxis (i.e., a state of hyper-susceptibility which is the reverse of immunity) do not invalidate the general rule that organisms respond to stimuli in a purposive manner. Berg also recognizes that organic characteristics which are neutral in relation to the life or death of the individual arise in a purely mechanical manner, although in obedience to physical and chemical laws and not by chance.[14] Berg's reference to the mechanical origin of existentially neutral characteristics recalls Aristotle's assertion that certain organic phenomena, such as the color of a person's eyes, are to be explained by material and efficient causes only, without the operation of final causality (*Generation of Animals*, 778a–b).

For Berg, the fundamental problem of evolution is to account for the origin of purposive adaptations. Starting with Empedocles, various thinkers over the centuries have suggested that the origin can be ascribed to a fortuitous combination of circumstances, which is the view taken and elaborated by Darwin with his theory of natural selection. According to this theory, purposive characters are the result of accidental usefulness, and the gradual perfection in organization, or progress, results from survival of the fittest. However, Berg objects that this fails to account for the *origin* of purposive characters, and only explains why individuals with useful characters survive and become more perfect. Let us look at an instance that Darwin himself mentions, the amphibian genus *Proteus*, which possesses both gills and lungs. When living in deep water, its gills grow

---

13. Ibid., 2.
14. Ibid., 1, 3.

at the expense of its lungs, while in shallow water the opposite occurs. Berg claims that Darwin's admission of the principles of use and disuse, and of adaptation to the environment, actually represents an implicit acknowledgement that primordial fitness is inherent in every living body, so that a theory of selection becomes superfluous.[15]

Is organic adaptation driven by accident or purpose? Against the Darwinian notion that sufficient time can produce almost infinite variation, Berg argues that the probability of an accidental origin of purposive contrivances does not increase with vast intervals of time, because time increases the number of both favorable and unfavorable variations. Darwin assumed that the variability of organisms is so great that chance, in adapting characters, will always select an accidental variation that may prove useful. Since for Darwin selection operates on accidental variations, every purposive adaptation must possess a very long history, during which natural selection weeds out those individuals that lack the necessary qualities. But in reality many adaptations are displayed spontaneously and immediately, without any intervention by natural selection. This is illustrated by Berg with examples from both the plant and animal kingdoms. One such is the phenomenon of graft hybrids, where we see that in the grafting of various pairs of plant species, for example tomato (*Solanum lycopersicum*) and nightshade (*S. nigrum*), groups of cells grow concordantly and produce stems, leaves and flowers consisting of tissues of both the scion and the stock. The same phenomenon occurs when parts of the embryos of the common newt (*Molge vulgaris*) and the crested newt (*M. cristatus*) are transplanted. It therefore appears as if an "inner regulating principle" had created hybrid forms out of parts of the organs of two different organisms. Similarly, experimental inter-species transplantations of body parts of animal types as diverse as newts, hydras, earth-worms, moths, and frogs have produced a single organism with characteristics of both species, which respond in a purposive manner to stimuli and in some cases have even reproduced. Berg concludes, "As in their former (phylogenetic) experience the individuals selected for

---

15. Ibid., 4–5, 14–15; Cooke et al., *Animals*, 423.

grafting or transplantation had never encountered anything of the kind, it obviously follows that the capacity for reacting in a purposive manner is commonly developed, not as a result of natural selection, but owing to that capacity being originally inherent in the organism."[16]

According to Immanuel Kant, the races of *Homo sapiens* are derived from a common genus through the influence of climatic conditions.[17] This implies that humans, like other animals, possess "an original predisposition to react in a different manner under varying climatic conditions."[18] Berg notes that this very predisposition includes a causal explanation, writing that "the admission of the principles of use and disuse, as also of adaptation to the environment, is a veiled acknowledgement that primordial fitness is inherent in every living body."[19] More recently, David Swift has accounted for variations in skin color among the human races in terms of genetic diversity and natural selection. He suggests that early *Homo sapiens* probably had more genetic variability, enabling the production of a range of skin colors. Worldwide migrations resulted in different gene combinations being selected in the different populations, suitable to their particular environment, and this resulted in a fair skin to those living in less sunny parts to facilitate the production of vitamin D, and a dark skin to those in more sunny parts as protection against ultraviolet light.[20]

Law, Chance, and Design

What is natural law? In nature, the same phenomena are repeated when the same conditions recur. Natural law is the probability of a future occurrence of such repetition.[21] Empirically speaking, natural laws pertain to numerical regularities within the realm of measurable phenomena. According to modern science, natural phe-

---

16. Berg, *Nomogenesis*, 36–38, 40, 42.
17. Ibid., 15, ft. 1; in our discussion, "race" has the scientific meaning of "subspecies" and not its unscientific conflation with nationality.
18. Ibid.
19. Ibid., 15.
20. Swift, *Evolution*, 254.
21. Berg, *Nomogenesis*, 133.

Evolution According to Natural Law

nomena should be explained by means of the simplest laws possible. These laws are viewed by many scientists as final and universal.[22] However, the assumption that nature always makes use of the simplest means is an *a priori* fallacy, as John Stuart Mill argues in his *System of Logic*.[23] Moreover, our knowledge of the laws of nature is by no means final, as the physicist David Bohm notes, writing, "the possibility is always open that there may exist an unlimited variety of additional properties, qualities, entities, systems, levels, etc. to which apply correspondingly new kinds of laws of nature."[24] Along related lines, Perry Marshall writes that randomness is always with respect to something. For example, when a raindrop falls on your windscreen it is random with respect to the motion of your car, but not with respect to the cloud it fell from. It is therefore conceivable that everything is determined by natural laws, with the only limitation being our ability to observe those laws.[25]

Lev Berg insists that "evolution is nomogenesis," by which he means that development is according to definite laws, and not due to chance, as Darwin believed.[26] That chance plays a subordinate role to law in the organic realm has more recently been affirmed by French zoologist Pierre Grassé. In his *Evolution of Living Organisms* (1977) Grassé writes, "It is neither randomness nor supernatural power, but laws which govern living things; to determine these laws is the aim and goal of science, which should have the final say."[27] And since our universe is subject to definable laws, it excludes anything that is incompatible with those laws. The law-regulated cosmos could thus be viewed as a system of mutually compatible possibilities which are not assembled by chance. Things exist because they are compatible with the conditions of that particular system, such as form, number, time, space and energy.[28]

Does the existence of chance refute causality? Absolutely not,

---

22. Thompson, *Science*, 134–35.
23. Berg, *Nomogenesis*, 20.
24. Quoted in Thompson, *Science*, 134–35.
25. Marshall, *Evolution*, 286–87.
26. Berg, *Nomogenesis*, xvii.
27. Quoted in Davison, *Hypothesis*, 5.
28. Northbourne, *Progress*, 113, 117.

Berg replies, if causality is understood as conformity to law. This argument reflects Aristotle's view that chance events are not anti-teleological in nature, but rather act as incidental causes in the attainment of purpose (*Physics*, Book II). Darwin himself acknowledged in *Origin of Species* that variations are only apparently due to chance, and are ascribed to the latter due to our ignorance of the particular causes. In a later work he added that each modification is subject to the law of causality and therefore not accidental.[29] When the philosopher Daniel Dennett writes that the design in nature recognized by Darwin is "a wonderful wedding of chance and necessity," he is committing an act of doublethink, since design as chance is not design, as Michael Flannery remarks.[30] Moreover, Dennett's insistence (following Darwin) that, given enough time, random copying errors will always produce an abundance of useful errors, is invalidated by information theory, especially information entropy. As Perry Marshall quips, "His [Dennett's] version of evolution is the information equivalent of a perpetual motion machine that gets faster and faster every year."[31]

In the pre-Darwinian era the proponents of natural theology based their arguments for design on the observed complexity of morphological structures such as the human eye. However, this teleological argument was undermined by the Darwinian notion that the phenotype is determined by the genotype. Genes are the source of useful variations which are then acted upon by natural selection in order to facilitate the adaptation of organisms to their mode of life. But, as David Swift points out, the recent discoveries of molecular biochemistry have undermined the Darwinian view that even macro-evolutionary changes can be explained in terms of variation and selection. For example, the substantial complexity of biological macromolecules (for example DNA) could not possibly have arisen through chance processes, let alone the integrated action of many such molecules in any organic system. Complexity is

---

29. Berg, *Nomogenesis*, 21–23; Darwin, *Origin*, 102.
30. Flannery, *Wallace*, 47.
31. Marshall, *Evolution*, 291.

Evolution According to Natural Law

therefore not incidental, but an essential aspect of molecular biology. Swift writes:

> So the argument for design, or purpose, or teleology in biology is weak at the morphological level. But at the molecular level we have an unequivocal case.... In summary, as originally formulated, even going back to Aristotle ... the argument for design was essentially subjective; but what I am suggesting here is an objective assessment based on the functionality and improbability of biological structures at the molecular level.[32]

Thus is the philosophical notion of teleology confirmed by evidence from molecular biology.

The immense complexity of the organic realm is evident even from one of the smallest and simplest living organisms, the unicellular bacterium *Escherichia coli*, which is commonly found in the intestines of warm-blooded animals. Although it is around 500 times smaller than the average cells of higher plants and animals, *E. coli* contains between 3000 and 6000 different types of molecules, as well as between 2000 and 3000 different kinds of proteins. Each of the latter is built up from 20 different amino acids, and a typical protein in an *E. coli* contains around 300 subunits of these amino acids.[33] Furthermore, not only does *E.coli* display an amazing structural complexity for such a minute organism, it also functions with astonishing accuracy. For example, its genome containing 4.6 megabytes of data is duplicated in around 40 minutes, with a copying precision of 99.9999999%. This is achieved through a three-step process of error correction, as described by Perry Marshall: (i) copying, in which one mistake is made in every 100,000 letters; (ii) proofreading, which removes the incorrect base and inserts the correct letter, thereby multiplying accuracy by between 100 and 1000; and (iii) mismatch repair, in which error-free DNA replaces DNA with erroneous letters, thereby multiplying copying accuracy by a further 100. Through this intricate process of error correction the unicellular bacterium duplicates its genome with an accuracy of

---

32. Swift, *Evolution*, 216–18.
33. Thompson, *Science*, 111–12.

one mistake per billion letters.[34] Clearly, this autonomous capacity of cells for the correction of copying errors contradicts the Neo-Darwinian insistence that the evolutionary process is driven by random copying errors.

The extreme improbability of even the simplest self-reproducing systems having arisen by chance has been mathematically demonstrated by Fred Hoyle and Chandra Wickramasinghe in their book *Evolution from Space* (1981). These scientists calculated the odds of *E. coli* being constructed through random interaction as $10^{40,000}$ to 1. The 4.5 billion years allowed by the estimated age of the Earth would therefore be much too brief to allow such an event to occur. The improbability of any biological macromolecules being constructed at random has also been convincingly argued by David Swift. For example, the chances of the small protein cytochrome *c* with its 104 amino acids being produced randomly has been calculated as 1 in 10^{40}, whereas the entire age of the universe (around 15 billion years) is less than 10^{18} seconds.[35]

Since the middle of the twentieth century the life sciences have become dominated by the molecular paradigm, in which the central dogma (as Francis Crick labeled it) is the flow of information.[36] It is now axiomatic in molecular biology that all the information required to specify a cell is contained in its DNA. In the Neo-Darwinian conception, this coded information cannot be changed except through random mutations. The biologist Jerry Coyne notes that in this view the term "random" means that "mutations occur regardless of whether they would be useful to the individual."[37] However, as we have just argued above, the enormous complexity of genetic information discounts the possibility of its having arisen by chance. If we again consider the very simple *E.coli* bacterium, it has been calculated that its single cell contains up to 5.5 million bits of genetic information. Microscopic studies have found that vertebrate cells are between 20 and 50 times more complex genetically than *E.*

---

34. Marshall, *Evolution*, 105.
35. Thompson, *Origins*, 35; Swift, *Evolution*, 136–38.
36. Hewlett, *Biology*, 176.
37. Quoted in Marshall, *Evolution*, 283.

coli, which means that they should contain coding for at least 20 to 50 times as many kinds of protein molecules as in *E. coli*. Furthermore, genetic "programming" entails an interacting system of structural and regulatory genes. Whereas a structural gene defines a specific structural element of a living organism, a regulative gene controls the timing and ordering in the expression of other genes. It is therefore improbable in the highest degree that complex organic structures could have arisen due to random mutations in either structural or regulatory genes.[38] Recent studies have in fact confirmed that chromosomal reorganizations do not occur randomly, as is held by Neo-Darwinism. Instead, it appears that specific regions within the mammalian genome have been predisposed to both micro- and macro-evolutionary rearrangements.[39]

We have already noted the ostensible victory of Darwinism over natural theology, which attributes the complex design found in living organisms to a transcendent intelligence. Since Darwin's principle of natural selection appears to explain the origins of biological form in terms of simple natural laws, it came to be widely accepted as scientifically superior to previous explanations of organic complexity. However, advances in molecular biology and probability theory have served to undermine the Darwinian theory. Thompson writes that

> the origin of complex order can be explained neither by natural selection nor by any other principle based on simple natural laws. Natural structures and patterns of high information content are inherently inexplicable by the reductionistic methods of quantitative analysis. The hypothesis that these structures have been generated by a transcendental intelligent being is therefore in no way inferior . . . as a theoretical explanation. In addition, this hypothesis opens up the possibility that if the ultimate foundation of reality is a sentient being, then it may be possible to acquire absolute knowledge directly from this transcendental source.[40]

---

38. Thompson, *Science*, 112–13, 198, 201.
39. Davison, *Hypothesis*, 4.
40. Thompson, *Science*, 135–36.

Nonetheless, it has long been objected that the numerous natural phenomena appearing as chaotic and meaningless cannot be reconciled with the traditional notion that all the phenomena in the universe arose due to the activity of a transcendent Intelligence. Richard Thompson provides an elegant response to this challenge, namely that meaningful patterns may appear random if they contain a high density of information. Since complex patterns tend to obey statistical laws merely as a consequence of their large information content, an apparently random pattern in nature may actually be meaningful. In addition, "meaningless patterns can easily be generated by the transformation of meaningful patterns. This is illustrated by the fact that many meaningful conversations, when heard simultaneously in a crowded room, merge together into a meaningless din." Ultimately, the concepts of meaning and purpose cannot be satisfactorily explained in terms of the mechanistic world-view, since they are transcendent in origin.[41]

The Role of Selection

Berg claims that the Darwinian theory of natural selection is a veiled acknowledgement of the principle of the inherent fitness of living beings, since it is based on the following principles which are purposive: heredity, variation, and the struggle for existence, for "if there were no variation, no organism could adapt itself to varying external conditions; if there were no heredity, it would be impossible to fix acquired characteristics; and, lastly, the struggle for existence postulates the faculty of self-preservation." The latter is synonymous with fit behavior, since organisms realize the minimum of variation and the maximum of heredity necessary for self-preservation.[42]

Natural and artificial selection both destroy useless forms, that is to say the unfit. However, in addition to "de-selection" of the unfit, evolution also requires selection of the fit, but this can only be achieved through artificial selection, which is the product of man's willed, intelligent actions. As Berg notes, "The creative side of selec-

---

41. Ibid., 164–65.
42. Berg, *Nomogenesis*, 18–19.

tion is found in artificial, not in natural selection." This is illustrated by the difference between the winter and spring varieties of cereal plants. In some cases the spring forms are derived from the winter forms, and in others *vice versa*. It was therefore not the artificial selection of *accidental* variations that caused the transformation of winter into spring forms or of spring into winter; instead, Berg writes, "Both the spring and the winter forms have existed in Nature from the very beginning. The role of selection is limited to the segregation of pure lines of spring races from mixtures of winter and spring forms, which are observed to occur among wild plants."[43]

It should be noted that new forms of domestic animals and cultivated plants can only be produced if a given species possesses the requisite tendencies, or, in metaphysical terms, if the inherent possibilities are already present in those organisms. These new forms are at any rate all derived from wild forms of which some already had highly diversified characters, such as the dog and the wheat. Therefore, Berg notes, in artificial selection the new is but a revelation of the old. This was in fact recognized by Darwin, when he wrote that man can only act by selection when variations are first given to him by nature.[44] In his critical assessment of Berg's theory of nomogenesis, the embryologist Ernest MacBride agrees that "natural selection is a purely negative agent: it weeds out, but it does not create: it accounts for the elimination of the unfit, but not for the appearance of the fit." He adds, "What the belief in natural selection as an efficient agent really implies is the constant occurrence of small inheritable variations in all directions. This assumption is directly contradicted by every relevant experiment devised to test the point. *If the conditions are kept constant* selection is powerless to effect progressive change."[45]

In support of Darwinism, Theodosius Dobzhansky writes that adaptive genetic changes brought about by natural selection cause

---

43. Ibid., 51, 54.
44. Berg, *Nomogenesis*, 57; Darwin, *Origin*, 32.
45. MacBride, E.W., "Berg's Nomogenesis. A criticism of natural selection," in *The Eugenics Review* 19(1), (1927): 31. http://www.ncbi.nlm.nih.gov/pmc/articles/PMC2984689/.

living species to respond to environmental challenges such as the Ice Age climate yielding to a warmer climate. If the response is successful, the species will occupy the new ecological niche; if unsuccessful, it will become extinct. He adds that "there is, of course, nothing conscious or intentional in the action of natural selection." It is rather a blind yet creative process, which has produced both *Homo sapiens* as the "apex of evolution" and organic forms of extreme over-specialization, such as a fungus of the genus *Laboulbenia*, which grows only on the rear portion of the elytra of the beetle *Aphenops cronei*, which in turn is found only in limestone caves in the south of France.[46] In the Neo-Darwinian view, organic diversity is attributable primarily if not exclusively to the unconscious yet creative power of natural selection.

What is the real scope of natural selection? Berg contends that it contributes to the geographical distribution of species and also "cuts off deviations from the standard by destroying extreme variations." Berg demonstrates the conservative tendency of natural selection using the phenomenon of mortality. For example, in the cultivation of beans, it is seeds of medium weight, not the lightest or heaviest, that survive best; and when a storm killed a large number of sparrows in a certain area, it was observed that the victims were those that differed most from the local sparrows in the size of their tails, wings, or bills. These observations serve to refute the principle of selection, since the struggle for survival should favor individuals that deviate most from the standard, with a variation that might by chance be useful. Berg argues that "if a useful character or contrivance is to have any chance of being preserved, it should be simultaneously manifested in a *vast number of individuals*; such individuals would be the most resistant to unfavorable conditions, and the extreme deviations would perish."[47] According to this argument, new species arise as a result of simultaneous mass mutations, as opposed to the Darwinian mechanism of extremely gradual change.

As we can see from the foregoing, Berg holds that natural selection is a conservative agency, acting to maintain the standard and

---

46. Dobzhansky, *Evolution*, 126.
47. Berg, *Nomogenesis*, 60–64, 370–72.

restrict variations, and thus cannot be the means of producing new forms. The French zoologist Pierre Grassé, too, asserts that selection acts more to conserve the inheritance of the species than to transform it.[48] Darwin himself admits that there are cases where all or most individuals of the same species have been similarly modified without the agency of selection, but he attributed it to a common inheritance producing similar variations. Berg counters that if a new form or variety occurs in a single individual or small number, it will in the due course of interbreeding dissolve in the general mass of parental forms, and adds that we may infer that a new form will become dominant only if it becomes so immediately upon appearing. The implication for Darwin's theory is inescapable: "Since the struggle for existence does not lead to the preservation of single favored individuals, but, on the contrary, tends to maintain the standard, all theories of evolution based on natural selection fall to the ground."[49]

From the point of view of information theory, for evolution to occur entropy has to be reversed, that is, adaptations have to produce increasingly higher levels of order. What complicates matters for natural selection is that it cannot select any unit smaller than one organism, since organisms compete as individuals, and even the smallest known organisms possess more than 300,000 base pairs, each containing two bits of information. Since entropy is a given, neither random mutation nor natural selection can increase the quality of genetic information. As Marshall notes, at best natural selection can reduce the speed of information entropy but can never reverse it. What this means in practical terms is that natural selection is able to slow down the extinction rate of a species undergoing detrimental mutations, but it cannot create a new species.[50]

Based on his critical survey of evolutionary theory in the light of biochemical, embryological and paleontological evidence, David Swift concludes that the notion of natural selection acting on randomly occurring variations does indeed explain a range of biologi-

48. Ibid., 400; Davison, *Manifesto*, 9.
49. Berg, *Nomogenesis*, 372–74, 402.
50. Marshall, *Evolution*, 298.

cal phenomena. First, by weeding out less fit individuals, natural selection maintains the general health of a species. This is its conservative role, as noted by Berg and others. Second, since the recombination of genes results in a random mixture, natural selection perpetuates the beneficial combinations and rejects the detrimental ones. It is this function that facilitates the adaptation of a species to changing conditions, as we have seen with melanism in peppered moths. Third, natural selection at least partly explains the co-adaptation of different species, as in mimicry and pollination. Finally, in some cases natural selection results in the refinement of characteristics, for example in the cheetah.[51] From this we conclude that Berg's relegation of the role of natural selection to the geographical distribution of a species does not do justice to its actual scope—at least as far as micro-evolution is concerned.

The Course of Evolution

Berg suggests that an organism experiences two kinds of effects in the process of its evolution: those inherent in its own organization and chemical properties, or autonomic (from the Greek *autos*, self); and those that arise from its geographical environment, or choronomic (from the Greek *choros*, place). These autonomic and choronomic causes are the two major laws which regulate the development of the organic world and thus determine the outcome of evolution.[52] The American geneticist Richard Lewontin argues along similar lines, writing, "A living organism at any moment in its life is the unique consequence of a developmental history that results from the interaction of and determination by internal and external forces."[53] The paleontologist Otto Schindewolf asserts that in evolution the internal is prior to the external, noting that the fossil record shows no evidence of a direct response by organisms to environmental influences and arguing instead that "the environment can only provoke and set in motion some potential that is

---

51. Swift, *Evolution*, 381–82.
52. Berg, *Nomogenesis*, 68–69.
53. Quoted in Van Vrekhem, *Evolution*, 198.

already present,"[54] or, to use Berg's terms, autonomic causes precede choronomic causes in the evolutionary process.

Berg claims that the phenomenon of phylogenetic acceleration proves that evolution is largely the unfolding of pre-existing rudiments, which is to say that evolution is to a considerable degree predetermined. Phylogenetic acceleration refers to the premature appearance of advanced features in so-called primitive organisms, such as a true placenta in the shark *Mustelus laevis*, the complex "organ systems" within a single cell of the ciliate protozoon *Diplodinium ecaudatum*, or the pneumatic bones in certain flightless reptiles. Features such as the latter are of no adaptive value, thereby suggesting the primacy of internal or autonomic factors in evolution. It is also noteworthy that *Diplodinium*, which live in vast numbers in the stomachs of cattle, possess structures similar to a brain, nerve connectives resembling those of annelids and arthropods, muscles, a kind of segmental spinal column, a mouth, esophagus, rectum and anus—astonishingly, all within the confines of a single cell. For American biologist John Davison, this suggests "that all the necessary information is already present for these structures at the protozoan level."[55]

Unlike Darwin, who attributes the extinction of organisms to natural selection, Berg claims that extinction is due to autonomic and choronomic causes. He writes, "Every group of organisms in the course of a definite period attains its optimum, after which, obeying certain impulses concealed in the constitution of the organism, it becomes extinct or is relegated to a secondary position, yielding its place to others," and offers as example the gymnosperms and reptiles that dominated in the Mesozoic era yielding in the Tertiary period to angiosperms and mammals.[56] As a matter of fact, it is acknowledged in modern biology that extinction is a natural aspect of evolution. At the background or normal extinction rate, a small number of species become extinct every year due to natural causes, while in the present mass extinction, which is sadly

---

54. Quoted in Davison, *Hypothesis*, 2.
55. Berg, *Nomogenesis*, 402; Davison, *Hypothesis*, 1–2; Davison, *Manifesto*, 40.
56. Darwin, *Origin*, 85–86; Berg, *Nomogenesis*, 70.

due to human activity, it is occurring at a faster rate than ever before.[57]

Berg adds that external causes contribute to the extermination of certain forms "only when this is so predetermined by internal agencies."[58] As mentioned earlier, this phenomenon supports the idea that the ability of organisms to adapt to their environment reveals the presence of a certain "inner regulating principle." In this way paleontology concurs with Aristotle's dictum that nature produces such things, which being impelled by a certain principle confined within them, attain a predetermined end. Davison agrees with Aristotle and Berg on the presence of auto-regulation in the evolutionary process. Noting that the vast majority of all the organisms that ever existed have become extinct, the American biologist suggests that they became extinct because they could no longer evolve or otherwise manage to survive.[59]

We know from the geological record that there have been at least six mass extinctions of life forms over the past 500 million years. The most devastating occurred around 250 million years ago at the end of the Permian period, when over 80% of all species on Earth were wiped out. It appears that the oxygen level dropped from 30% during the Carboniferous period to only 13% by the late Permian, which caused the extinction of most plant and animal life on Earth, even including trilobites. Another notable mass extinction, the so-called K-T event, occurred on the boundary between the Cretaceous and Tertiary periods around 65 million years ago. It was probably caused by a gigantic asteroid striking the Earth off the Yucatan coast, which brought about the extinction of dinosaurs and the eventual rise of mammals from small shrew-like forms to planetary dominance.[60] Not all scientists agree with the prevailing view that dinosaurs were exterminated in a sudden terminal event. The astrophysicist John Gribbin, for one, writes that paleontological evidence in North America indicates that dinosaurs gradually died out

---

57. Cooke et al., *Animals*, 26.
58. Berg, *Nomogenesis*, 71.
59. Ibid., 71–72; Davison, *Manifesto*, 8–9.
60. Alexander, *Creation*, 104–07.

Evolution According to Natural Law

over a period of at least 10 million years.[61] Be that as it may, external causes clearly do play a role in the evolutionary process, even if mostly in the form of mass extinctions, which in turn lead to vastly increased opportunities for the survivors.

Evidently nature strives in both plants and animals to insure in the best possible manner the preservation and welfare of developing progeny, Berg notes. That is why the embryo develops in the body of its mother, as we see in viviparous mammals and seed plants, i.e., gymnosperms and angiosperms. Evolution follows a definite course shaped by both external (geographical) and internal (autonomic) causes. Moreover, "Evolution bears a sweeping character, and is not due to single, accidentally favorable variations."[62] We contend that this statement pertains to macro-evolution and not to micro-evolution, which is effected mainly by variation and selection.

The Origin of Genes

John Davison argues that just as the information for organic evolution has been predetermined in the evolving genome, so the information required to produce a complete organism is contained within the fertilized egg cell. Both embryonic development and the evolutionary process involve the ordered de-repression of pre-existing information, so that "ontogeny and phylogeny become part of the same organic continuum utilizing similar mechanisms for their expression." As Richard Goldschmidt claims in his 1940 work *The Material Basis of Evolution* and as is confirmed in later studies, the chromosome and not the gene is the unit of evolutionary change. The primary differences between humans and the higher primates lie in the structural arrangements of their chromosomes, and not their DNA, which is to say that the specific information pertaining to humans, chimpanzees, orangutans, and gorillas was present in a latent state and revealed, or de-repressed, when their chromosome segments were placed in a new configuration. Since these new configurations or "position effects" do not involve the acquisition of new information from outside the genome, it is inconceivable that

---

61. Van Vrekhem, *Evolution*, 166–67.
62. Berg, *Nomogenesis*, 113–14, 400.

natural selection could have made any contribution to such chromosome reorganizations. Accordingly, Davison writes, "all genetic (evolutionary) changes originate in individual cells, in individual chromosomes, in a particular organism."[63]

However, as David Swift admits, at present we do not have a satisfactory naturalistic explanation for the origination of genes. He writes, "the information to produce (potentially) useful variations resides in genes, but we have no adequate explanation for how that information—meaningful nucleotide sequences (and the decoding mechanisms)—arose."[64] And since code cannot be produced by the laws of physics and chemistry, it is reasonable to conclude that genetic information originates in the activity of the intelligible reason-principles (Greek *logoi*) indwelling all existing things, so that we might say the genetic code is designed by a transcendent Intellect and is not due to a Neo-Darwinian chemical accident.

The Orthogenetic Formation of Characters

It has been observed that flying fish (genus *Exocoetus*) leap out of water and glide for several hundred meters without any motion of their elongated pectoral fins. On the other hand, various species of fish that exhibit a similar elongation of pectoral fin, such as the Aral Sea Roach, do not leave the water, while many other species of fish that leap out of the water to migrate or escape predators do not possess elongated fins. Evidently, the act of flight played no role in the formation of the fins. Berg concludes from this that elongated pectoral fins were formed because they were inevitable, which is to say that the particular character was formed due to certain natural laws, without any role played by natural selection or the use/disuse of parts.[65]

Further cases of orthogenetic formation of characters are not difficult to find. The African egg-eating snake (*Dasypeltis scabra*) is able to swallow a bird's egg up to three times its own diameter, and has glands supplying extra fluids to the mouth to lubricate the egg

---

63. Davison, *Hypothesis*, 1, 4; Davison, *Manifesto*, 17.
64. Swift, *Evolution*, 399.
65. Berg, *Nomogenesis*, 135–40.

shells. While the snake's teeth are rudimentary, the first thirty vertebrae penetrate the wall of the esophagus and protrude into its cavity, and these "esophageal teeth" crush the swallowed eggs, the contents passing into the snake's stomach and the shell being cast out through its mouth. Berg suggests that the anatomical structure of the egg-eating snake was not adapted to a particular function, but the other way round: the snake feeds instinctively in this unusual manner to utilize the peculiar structure of its anterior vertebrae.[66] Another notable instance is the reduction of the eye in cave-dwelling and subterranean animals, which conjecture might have attributed to their lack of use except that some of their relatives living in the open also possess reduced organs of vision. Berg argues that it is therefore not living in caves that caused the blindness, but rather that animals with a hereditary tendency to reduced sight can survive better in caves than in the open.[67]

Turning to the plant kingdom, we find complex contrivances for fertilization in orchids or for the capture of insects in insectivorous plants. However, among the former some blossoms repel rather than attract insects, whereas among the latter some species, for example those in richer soil, survive without any insect food. These exceptions are clearly incompatible with survival of the fittest.[68] Earlier, Hugo de Vries, a Dutch botanist, had argued that adaptation is not dependent on selection, but that the formation of characters appears to be autonomous. He writes: "Specific characters have evolved without any relation to their possible significance in the struggle for life.... The usual phrase, that species are adapted to their environment, should therefore be read inversely, stating that most species are now found to live under conditions fit for them.... We could say that in the long run species choose their best environment."[69]

Berg mentions further instances of characters with no apparent use for their possessors. This includes the enlarged teeth of dog

---

66. Cooke et al., *Animals*, 407; Berg, *Nomogenesis*, 142.
67. Berg, *Nomogenesis*, 143.
68. Ibid., 144–46.
69. Quoted in Berg, *Nomogenesis*, 147.

salmon and humpback salmon males when they ascend rivers for spawning; the large teeth of the larvae of the eel *Anguilla anguilla*, which never eats before metamorphosis; and the crossed tusks of the North American mammoth *Elephas columbi*, which were not only of no use to it, but could rather have led to its extinction. The conclusion to be drawn from this wide range of examples is that characters are formed by internal causes, independently of the degree of utility to the organism. What we can infer from this, Berg writes, is that "the external features of organisms are the embodiment of accordance with certain laws; these features do not arise by chance."[70]

The Role of the Geographical Landscape

Choronomic influences affect an organism in an imperative manner, compelling individuals to vary in a determined manner as far as the organization of the species permits, so that "with identical or similar geographical conditions, identical or similar results are bound to follow."[71] For example, the number of both species and varieties of European fresh-water fishes increases as we proceed from north to south, but variations in widely separated genera exhibit a tendency to develop in one direction, so that we see, for instance, the same decrease in the number of rays in the dorsal and anal fins, as well as changes in coloration and scale sizes. As it is highly im-probable that similar variations have arisen by chance, it appears that several groups of species living in similar conditions are subject to parallel variations, independent of the effects of selection, and the latter can only operate within these limits.[72] This notion of parallel variations suggests directionality in the evolutionary process, as we will consider in the next chapter.

A striking case of parallel adaptation is found in the extremities of mammals, reptiles and insects inhabiting sandy deserts, for improved locomotion in the sand. It has also been observed that many species, belonging to different families, of butterflies, snakes, mollusks, and beetles, among others, display similar coloration

---

70. Berg, *Nomogenesis*, 148–49, 262.
71. Ibid., 265.
72. Ibid., 265–69.

Evolution According to Natural Law

when living in the same region.[73] An example of the latter which is well known to animal lovers is the white coloration of various mammal and bird species living in the regions around the North Pole, such as the polar bear, the Alaskan gray wolf, the Arctic fox, the stoat, the Arctic and snowshoe hares, the Siberian collared lemming, the snowy owl, the snow goose, and the rock ptarmigan. Whereas the polar bear, as an apex predator, remains white throughout the year, most of the smaller species exhibit a darker coloration of fur or feathers as camouflage during the summer months.[74]

The plant kingdom, as well, displays numerous examples of similar development under the influence of similar external conditions. These include mountainous, polar, aquatic, xerophyte, and mangrove species. Berg considers the objection that organisms established themselves in a particular habitat because they were pre-adapted with suitable features, but since there are numerous recorded cases of plants directly adapting to a changed environment, he concludes, "We here perceive the direct effect of the geographical landscape, and there can be no application of the [Darwinian] principle of infinite variability, or of the selection of the accidentally best-adapted individuals: *all* individuals of a given landscape vary in a determined direction." That choronomic effects also occur among human beings is evident from anthropological studies showing that the descendants of European settlers in North America underwent similar variations in bodily characters, for example in the shape of the skull, due to the effects of the geographical landscape. These variations, Berg points out, affect all the individuals in the particular immigrant group.[75]

Another person to recognize the influence of the geographical landscape in the production of new forms is Dobzhansky. He distinguishes between two kinds of evolutionary changes: anagenesis, when a species undergoes adaptive changes in the face of environmental changes but remains a single species; and cladogenesis, when a species splits up into two or more species. The latter occurs mostly

---

73. Ibid., 282–83.
74. For illustrated examples, see Cooke et al., *Animals*, 127, 139, 227, 241, 254.
75. Berg, *Nomogenesis*, 271–73, 287–94.

when members of a species live in different territories with different environments. The separated species are first differentiated into races, which are then more and more genetically differentiated in response to their respective environments. Eventually, by means of natural selection, the genetic divergence is established as reproductive isolation. In this way the ancestral species is broken up into two or more new species. Dobzhansky argues that both anagenesis and cladogenesis occurred in the evolution of most plant and animal groups.[76] We contend that, understood in this way, anagenesis and cladogenesis can rather be viewed as stages of micro-evolution, since they do not require the generation of genetic novelty but merely the reshuffling of existing genes plus the action of natural selection to produce new races and eventually new species.

In a later work Dobzhansky employs the argument of adaptive radiation (i.e., the evolution of animals descended from a common ancestor to exploit different ecological niches that are unoccupied by other animals)[77] to explain the phenomenon of large numbers of species endemic to a small, isolated geographical area. He mentions the approximately 2000 species of the fruit fly genus *Drosophila*, of which around a quarter occur in Hawaii and nowhere else. Most of these Hawaiian endemic species are restricted to single islands or even to a part of an island (keeping in mind that the entire archipelago was formed relatively recently by volcanic activity and was never part of any continent). Dobzhansky attributes this phenomenon to a single species of *Drosphila* having arrived in Hawaii before any competitors, thus encountering many unoccupied ecological niches. The descendants of this first arrival responded to the environmental challenge by undergoing adaptive radiation, thereby producing the immense variety of *Drosophila* flies in Hawaii today, including the largest and smallest species of the genus.[78] However, the notion of adaptive radiation has been criticized as being an imposition of an evolutionary interpretation on the appearance of morphologically related groups, such as jawed fishes and flowering plants, in the

---

76. Dobzhansky, *Biology*, 123–24.
77. Cooke et al., *Animals*, 588.
78. Dobzhansky, *Evolution*, 128.

absence of common predecessors. The retropolation of adaptive radiation at lower taxonomic levels to higher levels could therefore be erroneous.[79]

The Polyphyletic Origins of Similar Forms

Darwin argued in *Origin of Species* that all animals are descended from a maximum of four or five progenitors, and plants from an equal or lesser number. He added, "Therefore I should infer from analogy that probably all the organic beings which have ever lived on this earth have descended from some one primordial form, into which life was first breathed." The latter view came to be known as the theory of monophyletic origin, emphasized by contemporary Darwinists such as Richard Dawkins, whose lack of philosophical and theological training does not deter him from often making doctrinaire pronouncements in these areas of knowledge. He confidently declares:

> We can be very sure there really is a single concestor [i.e., common ancestor] of all surviving life forms on this planet.... The evidence is that all that have ever been examined share ... the same genetic code.... As things stand, it appears that all known life forms can be traced to a single ancestor which lived more than 3 billion years ago.[80]

The monophyletic descent postulated by Darwin and his followers was supported from a different angle by the German zoologist Karl Schneider, who argued that every simple form potentially contains within it a more complex form, the latter evolving from the former. According to Schneider, "there must exist some one primordial form, within which all the remaining forms are 'enfolded' (*eingewickelt*), all the morphological potentialities having already been contained in the primordial cell (*Urzelle*)."[81] Ironically, while supporting monophyletism, Schneider also cited the traditional

---

79. Swift, *Evolution*, 295.
80. Darwin, *Origin*, 364; Dawkins, Richard, *The Ancestor's Tale. A Pilgrimage to the Dawn of Life* (London: Weidenfeld & Nicolson, 2004), 12–13.
81. Quoted in Berg, *Nomogenesis*, 358.

conception of evolution as the unfolding of inherent possibilities. This notion was supported by the geneticist William Bateson, who viewed the entire evolutionary process as "an unpacking of an original complex which contained within itself the whole range of diversity which living things present."[82]

Opposing the theory of monophyletic descent, Berg argues that due to the vast number and variety of life-forms on Earth the number of primal ancestors must be thousands or tens of thousands. Berg bases his theory of polyphyletic origins on a wide range of paleontological evidence. For instance, in the case of the dinosaur groups Saurischia and Ornithischia, the latter is not derived from the former as was earlier assumed, but instead they were parallel branches. In the phylum Arthropoda, the crustaceans and arachnids owe their origin to mesomeric worms, while the myriapods and insects owe their origin to polymeric worms. In the plant kingdom the angiosperms and gymnosperms developed independently from different ancestors. Furthermore, organs such as the notochord, gill-slit, and dorsal nervous system developed independently in various groups of animals, and not only in the ancestors of fishes. As further examples from paleontology of parallel, independent development of characters, Berg includes molluscs, ammonites, and flightless birds. The principle of polyphyletic origins leads to the conclusion that "points of resemblance in two forms may represent something secondary, acquired and new, whereas points of dissimilarity are something primary, inherited and ancient." This morphological convergence is the opposite of Darwin's law of divergence, which Berg does not deny, but he holds that the law of convergence operates alongside it and even dominates it.[83]

Another argument Berg uses for polyphyletic origins is the phenomenon of polytopical formation, or the appearance of identical forms in various places within the limits of one species, which forms are hereditary as long as the conditions remain invariable. It is evident, for example, in a number of marine salmon and trout

---

82. Ibid., 359.
83. Ibid., 338–39, 341, 345–47, 358.

species found across the northern hemisphere, of which some species remain permanently in fresh water where they give rise to dwarf forms. This is also an instance of evolutionary convergence, Berg notes, inasmuch as it occurs among various species. The independent origin of the same form in different places has also been observed in various species of plants and invertebrates.[84]

The French geologist Paul Lemoine points out that the geological record indicates the sudden appearance of all new groups, whereas changes in existing animals take place slowly, as a rule. Using the evolutionary rates of insects as an example, he argues that belief in the monophyletic origin of life requires invoking astronomical periods dating back to times when the Earth or even the solar system had not yet come into existence.[85] The theory of polyphyletic origins is moreover consonant with the observed absence of transitional forms between separate groups, since there is no common root. Berg concludes, "Both convergence and the absence of transitions support the view that evolution advances *by means of the transformation of vast numbers of individuals into new forms.*"[86] The fascinating phenomenon of convergent evolution will be discussed in a later chapter.

The Formation of New Species

Regarding the formation of new species, or speciation, Berg asserts that the production of new forms occurs in the context of geographical isolation. Opposing Darwin's view that only individual variations contribute to the production of new species, natural selection acting on only a few inhabitants of the same region, he writes: "In the origination of new geographical forms...a vast number of individuals inhabiting a certain geographical area are simultaneously involved in the production of new characters. There can be no question of an accidental occurrence of characters. Thus, in the production of geographical forms natural selection plays no

---

84. Ibid., 351–56.
85. Dewar, *Illusion*, 261.
86. Berg, *Nomogenesis*, 361 (italics in the original).

part." Instead, geographical forms are apparently the result of external agencies such as the landscape (as studies of animal groups as widely diverse as insects, fishes and rodents suggest), whereas non-geographical forms are due to internal causes. In the formation of species, Berg concludes that there is "the simultaneous *en masse* manifestation of new characters over a vast territory." This contradicts Darwin's view that natural selection acts very slowly and on only a few of the inhabitants of the same region at the same time, though we should note that in his later work, *Variation of Animals and Plants*, the English naturalist admitted that new sub-species could be produced through external conditions without the aid of selection.[87]

A new species is formed either by divergence or substitution. In the case of the former, a sub-species separates from its species, whereas in the latter, a younger species is substituted for a maternal one through mutation. In both cases the change is effected by a transformation *en masse*. A species occupies a definite territory, since its formation entails the simultaneous transformation into a new form of a vast number of individuals over a relatively extensive, continuous territory. Every taxonomic unit is above all characterized by morphology, yet morphology is itself the result of both inner and external causes.[88]

In discussing the mass appearance of similar variations in the living kingdoms, Berg uses the analogy of new formations in language. The latter arise simultaneously in large groups of individuals, while single variations disappear with the death of the individuals using them. Moreover, linguistic changes occur simultaneously over large territories, of which examples can be found in Russian, Spanish, French, and Classical Greek. According to studies by the French linguist Antoine Meillet, innovations occur among all children born in the same locality separate from the speech of the adults. Such variations are transmitted to younger generations, modifying the language over time. As with animals and plants, Berg contends, so also

---

87. Darwin, *Origin*, 84; Berg, *Nomogenesis*, 363–66, 368–69.
88. Berg, *Nomogenesis*, 396–99.

in languages new formations rapidly succeed each other in some complexes, i.e., species or languages, while in others characters are more stable.[89]

In contemporary biological terms, speciation occurs either through allopatric or sympatric mechanisms, which refer to geographical and reproductive isolation, respectively. Among plants, sympatric speciation is more common, whereas in animals, allopatric speciation occurs more often. The latter process occurs when a breeding population is split in two, as for example through continental drift. Both mechanisms have been observed to facilitate rapid speciation, as the molecular biologist Denis Alexander acknowledges, though he works within a Neo-Darwinian framework. Rapid speciation in the plant kingdom takes place by means of chromosomal doubling, also known as polyploidy. A prominent example thereof is the salsify genus *Tragopogon*, in which several new species have been formed in North America within a few decades.[90] This phenomenon could be viewed as an instance of micro-evolution, resulting from genetic reshuffling rather than mutations.

An example of rapid speciation in the animal kingdom is the cichlid fishes in the Great Lakes of Africa, of which more than a thousand new species have appeared during the past million years. More than 170 cichlid species are still living in Lake Victoria, filling a wide range of ecological niches and with the morphology of each species adapted to their feeding habits. At least five endemic cichlid species have evolved in the nearby and much smaller Lake Nagubago, which became separated from Lake Victoria around 4000 years ago. The remarkable case of the African cichlids shows that relatively rapid speciation can be stimulated by a suitable environment, Alexander notes.[91] The Lake Nagubago case was also commented on by the Neo-Darwinist Ernst Mayr, who contrasted it with that of the Panama Isthmus, which has separated the Pacific

---

89. Ibid., 374–77.
90. Alexander, *Creation*, 94–95.
91. Ibid., 98–99.

and the Caribbean for approximately 5 million years, yet the marine fauna with their massive gene pools on either side are virtually identical. As David Swift observes, these anomalous rates of evolution present a challenge to the Darwinian model, according to which a larger population will have an increased chance of favorable mutations, and will hence display more variability and a more rapid rate of evolution.[92]

The Limits Between Species

Naturalists borrowed the notion of species from philosophy, where it was introduced by Aristotle, who viewed a species as "an irreducible peculiarity which makes the thing what it is."[93] As we pointed out in an earlier chapter, for Aristotle, the essence of a living being is related to reproduction, so that a species is characterized by the ability to produce fertile offspring, in contrast to the reproductive sterility observed in hybrids.[94] It should be noted, however, that in some animal and plant species it is possible to produce offspring which possess new genetic features and, by reproduction, then create a new species. For instance, among salmon and salamanders a genome doubling occurs when different species mate. The offspring inherit a full set of chromosomes from each parent, which is called diploidy. In rare occasions the parental DNA strands are not only joined together but also rearranged (called hybrid dysgenesis), resulting in new features not possessed by any of the parents. For example, the wheat used for baking bread was obtained through blending emmer wheat with a noxious weed called goat grass.[95] These exceptions, nonetheless, do not invalidate the rule of hybrid sterility.

For Aristotle, moreover, the reproductive ability of a species is metaphysically based: "Since it is impossible for it [i.e., a living thing] to be eternal as an individual . . . it is possible for it to be eternal in species. This is the reason why there exists eternally the class of human beings, animals, and plants" (*Generation of Animals* 731b–

---

92. Swift, *Evolution*, 247–48.
93. Quoted in Berg, *Nomogenesis*, 389–90.
94. O'Rourke, *Metaphysics*, 25.
95. Marshall, *Evolution*, 136–37.

732a).⁹⁶ The Aristotelian view was cited by Buffon, who saw a species as a continuous succession of similar and multiplying beings. And according to Kant in his *Critique of Pure Reason*, species in nature are isolated from one another and thus form a discrete quantity.⁹⁷

Titus Burckhardt related the reality of species to the Hellenic notion of hylomorphism (from the Greek *hylē*, matter and *morphē*, form), which teaches that every individual thing is a conjunction of a specific matter with a specific form. While the form of a thing is the seal of its essential unity, its matter is "the plastic substance that receives this seal while conferring on it a concrete and limited existence." Therefore, "it is only in connection with a 'matter,' or plastic substance, that 'form' plays the part of a principle of individuation; in itself, in its ontological basis, it is not an individual reality but an archetype, and as such it lies beyond limitations and beyond change. Thus a species is an archetype, and if it is only manifested by the individuals belonging to it, it is nonetheless as real and indeed incomparably more real than they are." Moreover, since a species manifests an immutable "form," it cannot evolve and become transformed into another species. It can, however, produce variations as "projections" of a single essential form, like the branches of a tree.⁹⁸

Opposing the traditional concept of species, Darwin argues that the term species is arbitrarily given for the sake of convenience. He adds that a well-marked variety may justly be called an incipient species, and that therefore varieties have the same general characters as species, which is to say that the term "species" does not differ essentially from the term "variety," likewise applied arbitrarily for the sake of convenience.⁹⁹ This is reminiscent of the medieval debate between nominalists ("things having the same name share nothing more than the name") and realists ("things exist in reality, independently of our concepts and language"). It is our contention that the Darwinian rejection of the reality of species is linked to its insistence on gradualism: if minute variations indeed accrue incre-

---

96. Quoted in Gerson, *Aristotle*, 118.
97. Berg, *Nomogenesis*, 390.
98. Burckhardt, *Cosmology*, 140–41.
99. Darwin, *Origin*, 42, 47.

mentally over numerous generations to form new species, then there could *ipso facto* be no real boundaries between species.

In his book *The Ancestor's Tale*, Richard Dawkins goes so far as to attribute the notion of species to "the tyranny of the discontinuous mind," with its "obsession with discrete names." After thus displaying his nominalist proclivity, Dawkins admits that the majority of living animals are in fact discontinuous from one another, for example humans and chimpanzees. The English biologist attempts to explain this discontinuity by suggesting that the intermediates between surviving species are mostly extinct, so that "evolution tells us there are other lines of gradual continuity linking literally every species to every other." Following the example of Ernst Mayr, Dawkins lays the "delusion of discontinuity" at the door of Plato's philosophy of "essentialism" (which term, as we have shown in an earlier chapter, involves a misunderstanding of Plato's ontology). In reality, Dawkins continues, species do not change into other species but have evolved from a common ancestor. Echoing Darwin, he then suggests that we continue using names for apparently discontinuous species, while keeping in mind that they are no more than a "convenient fiction."[100]

Against the nominalists, Berg states that the notion of species is empirically confirmed. When a group of individuals resembling each other more than members of any other group is compared with another such group, and the differences between the two groups are sufficiently perceptible, we are dealing with a species. That is to say, species are those complexes of forms which are morphologically distinguished and also genetically different from neighboring form-complexes. Moreover, since mutation always involves leaps or interruptions, species are clearly distinguishable from one another. For their part, sub-species (or races) are always connected with species by means of transitions, but may in the course of time be disconnected from them either when a sub-species evolves into a new form, or when transitional forms become extinct.[101]

We should note however that in practice the definition of a spe-

---

100. Dawkins, *Ancestor*, 252, 258–59, 261.
101. Berg, *Nomogenesis*, 390–92, 395.

cies is not always a clear-cut matter. On the one hand there are instances, which although morphologically highly different, readily interbreed. On the other hand there are cases which are morphologically highly similar, but unable to interbreed. The latter, called "sibling species," were first discovered in the fruit fly *Drosophila*. From this phenomenon David Swift concludes that there are genetic isolating mechanisms which are unrelated to morphological differences.[102]

Mutation and Speciation

In 1869, following an exhaustive study of ammonites, the German paleontologist Wilhelm Waagen published *Die Formenreihe des Ammonites subradiatus*, in which he established mutations as the underlying cause of the evolution of species. Berg contends that these mutations arise due to a law inherent in the organism, to autonomic rather than choronomic causes. Since the mutational production of forms proceeds by sudden large-scale mutation, or saltation, the fossil record shows periods when nature displays a kaleidoscope of organic forms, and others when nature works at a slower rate, as would later be recognized in the theory of punctuated equilibrium. Berg adds that the mutational theory explains the phenomena of polyphyletic origins, the sudden appearance of species, and the absence of transitions between species. We could therefore say that species arise either through mutational transformation *en masse*, or through geographical isolation.[103] These two alternatives correspond to the distinction we are making between macro- and micro-evolution.

According to the mutational theory of the Dutch botanist Hugo de Vries, new forms are produced in sudden leaps. For example, the single-leaved strawberry *Fragaria monophylla* was obtained suddenly from seeds of the common European strawberry *F. vesca*, and proved to be constant. However, mutations are generally displayed in a very small number of individuals, so that in wild plants they usually become extinct, and in cultivation they often do not blos-

---

102. Swift, *Evolution*, 112–13.
103. Berg, *Nomogenesis*, 384–86.

som or produce seeds. The experiments by De Vries and others show that new forms cannot arise as a result of individual variations, whether hereditary or non-hereditary and Berg reiterates that for the production of new forms a transformation *en masse* of the entire or the major part of the individuals living in a definite territory is required, and that only in this manner can variation be fixed by heredity. As mentioned earlier, such a mass transformation of forms occurs due to either choronomic or autonomic causes.[104] Berg's theory of speciation through sudden mass mutation was criticized by Ernest MacBride for ignoring well-known fossil series such as those of the horse and the camel, and the imperfection of the fossil record.[105] We will consider the fossil record later in this chapter, while the fossil series can be considered as cases of microevolution due to gene reshuffling and natural selection.

Against Leibniz's dictum that *natura non facit saltus* (Latin, "nature does not jump"), Berg argues that rhythm is a manifestation of the law of intermittent development. Accordingly, "the birth or death of individuals, species, ideas is a catastrophic process. The manifestation of every class of these phenomena is preceded by a long latent period of development, which follows certain definite laws, and then suddenly culminates in a bound, *saltus*, by which the group emerges into the light, is distributed over the earth's surface, and wins for itself 'a place in the sun.'" This phenomenon that 'nature does jump' (*natura facit saltus*) is paralleled in quantum physics when energy emanating from a source is produced in packets and not continuously. Moreover, paleontology confirms the rapid production of new forms. Thus in the Permian period Stegocephala flourish, in the Triassic, reptiles, in the Lower Cretaceous, angiosperms, on the border between the Lower and Upper Cretaceous, teleostean fishes dominate, and in the Lower Tertiary, mammals. This evidence also serves to rebut Darwin's assertion that there are no sudden leaps in nature, since natural selection works slowly on slight, successive variations.[106]

---

104. Ibid., 377–78, 383–84.
105. MacBride, *Berg*, 37.
106. Berg, *Nomogenesis*, 387–89; Darwin, *Origin*, 355.

Evolution According to Natural Law

Berg's position on the sudden appearance of new species would surely be supported by the geneticist Richard Goldschmidt and the paleontologist Otto Schindewolf. Goldschmidt notes that micro-evolutionary adaptations to the environment do not exceed the boundaries of species, while in contrast, species and the higher categories originate in single macro-evolutionary steps as completely new genetic systems. These new systems are produced by means of a comprehensive transformation of intra-chromosomal structures, so that "a single modification of an embryonic character produced in this way would then regulate a whole series of related ontogenetic processes, leading to a completely new developmental type."[107] According to Schindewolf, macro-evolution takes place in an "explosive" way within a short geological time, followed by a slower series of orthogenetic perfections (i.e., micro-evolution). Schindewolf employs examples from the fossil record to show that the major evolutionary advances must have taken place in single large steps, "which affected early embryonic stages with the automatic consequence of reconstruction of all the later phases of development."[108] Interestingly, Schindewolf's view is reflected to some extent in the modern evolutionary notion of a two-phase process (German *Zweiphasenhypothese*), in which micro-evolutionary phases alternate with "creative bursts" through which fundamentally new forms are brought into existence.[109] These arguments from genetics and paleontology serve to reinforce our thesis of a fundamental distinction between macro- and micro-evolution.

The Laws of Evolution

Employing Max Planck's distinction between statistical and dynamic laws, Berg reasons that if evolution had taken place according to tychogenesis (from the Greek *tychē*, luck), its laws would have been similar to those physical laws which are of a statistical character. These are irreversible processes, such as the conductivity of heat and electricity, diffusion, friction, and chemical reactions—all of

---

107. Quoted in *Davison* 2000, 29–30.
108. Ibid., 30.
109. Smith, *Cosmos*, 89–90.

which are governed by the second law of thermodynamics, or entropy. In contrast, the laws of nomogenesis are of a dynamic character, namely reversible processes such as gravitation, electric and mechanical oscillations, acoustic and electro-magnetic waves.[110] Interestingly, this distinction implies that Darwinian evolution, or tychogenesis, is more static in nature than evolution according to natural law, or nomogenesis.

The following natural laws of evolution are recognized by Berg:[111]

(i) "Higher characters or their rudiments appear in lower groups very much earlier than they are manifested in full development in organisms occupying a higher position in the system"—evolution is therefore mostly the unfolding of pre-existing rudiments;

(ii) "The successive manifestations of new characters are governed by law. In the process of evolution there is no place for chance: new characters appear where they should appear"—both ontogeny and evolution are thus prescribed by law;

(iii) "Therefore evolution follows a determined direction"—instead of chaotic variation;

(iv) "Some characters owe their development to internal (autonomic) causes inherent in the very nature of the organism, and independent of any effects of the environment"—this is particularly evident in the ontogenetic process;

(v) "The laws of development of the organic world are the same both in ontogeny and phylogeny"—thereby affirming the "recapitulation" of phylogeny by ontogeny;

(vi) "Both in phylogeny and in ontogeny characters develop at a different rate: some repeat, as it were, the former stages, others predetermine the future ones"; and

(vii) "Every organism consists of a combination of characters which evolve to a considerable degree (sometimes entirely) independently one of another."

In summary, for Berg evolution is by and large autonomic and not ruled by chance. The formation of new characters also comes about

---

110. Berg, *Nomogenesis*, 405.
111. Ibid., 154–55, 234.

as the result of the geographical landscape, and may be termed choronomic in origin. Both ontogenetic and phylogenetic development follow certain laws common to both, thus confirming the reality of nomogenesis. This accordance with autonomic laws is best shown in the life-forms that developed convergently.[112] This latter phenomenon will be explored in a later chapter.

Darwinism *versus* Nomogenesis

In the closing pages of his monumental work, Berg juxtaposes the salient features of Darwinism on the one hand and nomogenesis on the other.[113] The Darwinian theory of evolution holds that: "(i) All organisms have developed from one or a few primary forms, i.e., in a mono- or oligo-phyletic manner. (ii) Subsequent evolution was divergent, (iii) based on chance variations, (iv) to which single and solitary individuals are subject, (v) by means of slow, scarcely perceptible, continuous variations. (vi) Hereditary variations are numerous, and they develop in all directions. (vii) The struggle for existence and natural selection are progressive agencies. (viii) Species arising through divergence are connected by transitions. (ix) Evolution implies the formation of new characters. (x) The extinction of organisms is due to external causes, the struggle for existence and the survival of the fittest."

In contrast, nomogenesis holds that: "(i) Organisms have developed from tens of thousands of primary forms, i.e., polyphyletically. (ii) Subsequent evolution was chiefly convergent (partly divergent), (iii) based upon laws, (iv) affecting a vast number of individuals throughout an extensive territory, (v) by leaps, paroxysms, or mutations. (vi) Hereditary variations are restricted in number, and they develop in a determined direction. (vii) The struggle for existence and natural selection are not progressive agencies, but being, on the contrary, conservative, maintain the standard. (viii) Species arising through mutations are sharply distinguished one from another. (ix) Evolution is in a great measure an

---

112. Ibid., 402–04.
113. Ibid., 406–07.

unfolding of pre-existing rudiments. (x) The extinction of organisms is due to inner (autonomic) and external (choronomic) causes."

Although Lev Berg appears to be right that the evolutionary process is determined by law at least insofar as it pertains to macro-evolution, it would be wrong to reduce the whole of the organic realm to the operation of physical and chemical laws. The renowned physicist Erwin Schrödinger himself argued that the structure of living matter cannot be reduced to the operation of the ordinary laws of physics. After all, the organic realm differs not only in degree, but also in kind from the inorganic. This was recognized by Schrödinger when he applied "order from order" and "order from disorder" to the organic and inorganic realms, respectively. He furthermore coined the term "negative entropy" when describing the ability of living things to regulate events. Nonetheless, as Wolfgang Smith points out, it is undeniable that there are physical laws operative within the biosphere, which explains the far greater prevalence of order in the living kingdoms than in the inanimate world.[114]

Ultimately, as Perry Marshall notes, we should recognize that although living beings obey the universal laws of physics and chemistry, these laws are insufficient to explain the behavior of living things completely, since living beings are subjective selves displaying intentional behavior. Nevertheless, since all life is based on codes such as the genetic code in DNA and each layer of code infers a higher level of intentionality, the behavior of organisms is code-guided and thus teleological.[115] In other words, an evolutionary process determined by law accommodates subjectivity while establishing purpose in the organic realms.

Paleontological Confirmation of Mass Mutations

Berg's postulation of simultaneous mass mutations as the primary means in the production of new species is strikingly confirmed by the so-called Cambrian Explosion. This event has been labeled the Big Bang of animal evolution, a "quantum leap of unprecedented

---

114. Marshall, *Evolution*, 219; Smith, *Cosmos*, 85–86.
115. Marshall, *Evolution*, 216.

Evolution According to Natural Law

proportions."[116] It refers to the sudden appearance in the fossil record during the Cambrian period (around 542 to 488 million years ago, hereafter MYA) of around 90 percent of the animal phyla living today. In the vast pre-Cambrian era the earliest microfossils date back around 3.5 billion years, to the Archaean. For the next three billion years prior to the Cambrian, life on Earth was confined to single-cellular and tiny multi-cellular forms, less than 1 mm in diameter and dominated by algae. Then, within a few million years beginning around 525 MYA, the ancestors of today's sponges, jellyfishes, corals, brachiopods, crustaceans, annelids, gastropods, cephalopods, sea lilies, and jawless fishes all appear on the scene, or rather in the rocks. This organic kaleidoscope includes complex animals with limbs, antennae, eyes, and tails. The Cambrian Explosion also witnessed the first appearance of animals with hard shells and exoskeletons, probably spurred by an increase in predatory species. The jawless fishes known as Agnathans appearing towards the close of the Cambrian, around 480 MYA, were the first vertebrates,[117] but to compound the Neo-Darwinian dilemma, these jawless fishes appear in large diversity as fully differentiated forms, with no evidence of intermediates linking them.[118]

Already by the middle of the twentieth century more than 1100 genera of Cambrian animal fossils found in numerous localities in both hemispheres had been described. All of these forms are said to have developed out of mostly single-celled ancestors. But, as Douglas Dewar argued, if the Darwinian hypothesis of slow, gradual speciation was correct, then the rocks of the pre-Cambrian periods should have yielded a plethora of transitional forms for this huge diversity of marine fauna. However, such evidence is lacking.[119] Exceptions are the Ediacaran fossils discovered first in southern Australia and then on other continents. These fossils, dated to around 565 million years ago, consist of soft-bodied marine fauna such as jellyfishes and worms, as well as forms unlike those of any

---

116. Alexander, *Creation*, 124; Marshall, *Evolution*, 134.
117. Cooke et al., *Animals*, 24, 27–28; Alexander, *Creation*, 88–90, 124.
118. Swift, *Evolution*, 264.
119. Dewar, *Illusion*, 19–20, 25.

other period. However, they became extinct in the early Cambrian and therefore played no further part in the evolutionary process. In addition, recent discoveries have narrowed the gap between the Cambrian and Ediacaran fossils from as much as 100 million years to 13 million years.[120]

Various explanations have been offered to account for the relative lack of pre-Cambrian fossils, presenting as it does a serious challenge to Darwinian gradualism. For instance, it has been suggested that all the pre-Cambrian rocks have been altered to the extent that the fossils they originally contained have all been destroyed. However, as Dewar points out, unaltered sedimentary rocks predating the Cambrian exist in many parts of the world, in layers several kilometers thick. Some of these are so perfectly preserved that fossilized water marks can be seen, but there is no evidence of organic fossils. Another transformist argument is that all the pre-Cambrian marine animals either lacked shells or had fragile shells that could not be fossilized. Again, this does not explain the existence of mollusks and brachiopods in the Lower Cambrian possessing thick shells, or the fact that soft-bodied animals such as jellyfishes have left perfect impressions in sedimentary rocks.[121] Moreover, David Swift writes, even if it is conceded that the soft-bodied predecessors of the Cambrian fauna were unable to leave any trace, so that we should not be surprised by the sudden appearance of one evolving line in the fossil record, this does not explain the sudden and simultaneous appearance of such a wide variety of forms with hard parts. Equally remarkable are the radically different skeletons of the trilobites, sponges, echinoderms, mollusks, and the first vertebrates appearing in the Cambrian. These skeletons are produced by intricate biological processes, which makes their simultaneous appearance even more difficult to explain in Neo-Darwinian terms.[122]

Remarkably, plant life remained mostly confined to algae for another 100 million years or so after the Cambrian period, until the Devonian (around 416–359 MYA). In Devonian rocks an abundance

---

120. Cooke et al., *Animals*, 27; Alexander, *Creation*, 122; Swift, *Evolution*, 260.
121. Dewar, *Illusion*, 27–28.
122. Swift, *Evolution*, 260–62.

of plants, at least ten classes, including fungi, club-mosses, ferns, horse-tails and Gymnosperms such as conifers, then make their appearance. These "higher" plants had complete vascular systems for the transport of water and nutrients, and again we encounter highly differentiated structures appearing within a relatively brief period in the distinct sub-phyla of ferns, horse-tails and club-mosses. With these vascular plants comes the appearance of the first leaves in two differentiated types: the "simple" leaf and the frond-like leaf as found in ferns, with no intermediates linking them or any known predecessors.[123]

Also making their first appearance in the Devonian is the class of Insecta. With around a million described species in 29 orders, insects comprise more than half of all known animal species, and it is estimated that several more million insect species are awaiting discovery. Inhabiting all of the Earth's habitats in prodigious numbers, insects are by various measures (quantitative ones, we should add) the most successful animals to have lived on Earth. The earliest fossil insects were wingless springtails and bristletails, both appearing fully developed with the tri-segmented body of modern insects. They were followed much later, in the late Carboniferous, by winged insects, which also appear fully specialized and with no known intermediates linking them with wingless insects. Among winged insects, both those with folded wings and those with rigid wings appear in the fossil record with no intermediates linking them to each other or to any predecessor. David Swift concludes: "Although the insects are the most diverse animal group, and have been abundant at least since the Carboniferous period, we cannot trace an evolutionary [understood in Darwinian terms of variation and selection] origin of insects as a whole or of their principal groups."[124]

Nor are these Cambrian and Devonian "explosions" the only instances of a sudden outburst of new life-forms found in the geological record. The following examples from various periods follow-

---

123. Cooke et al., *Animals*, 24; Dewar, *Illusion*, 40; Swift, *Evolution*, 277–78.
124. Cooke et al., *Animals*, 552; Swift, *Evolution*, 276–77.

ing the Cambrian are presented by Douglas Dewar[125] and expanded by other authors as mentioned:

(i) Ordovician (488–444 MYA)—two orders of jawless fishes, two classes of Echinoderms, 14 families of Crinoidae and 19 families of Bryozoa; also fossilized plant spores dated 475 MYA, and the first appearance of cartilaginous fishes, such as sharks;[126]

(ii) Silurian (444–416 MYA)—a sub-class of fish with jaws (Selachians), a class of Echinoderms, two orders of Echinoidae and two genera of scorpions; these arthropods were the first animals on land, their hard exoskeleton restricting water loss; also the first plants without leaves, with fossils dated 430 MYA;[127]

(iii) Devonian (416–359 MYA)—various bony fish groups, and the first appearance of millipedes, crustaceans, mollusks, and amphibians; due to the proliferation of fishes this period is often referred to as the Age of Fishes;[128]

(iv) Carboniferous (359–299 MYA)—12 orders of arachnids (including the first appearance of spiders), 12 orders of insects (of which most have become extinct), 14 families of amphibians, and the first appearance of reptiles; spiders thus appear simultaneously with much of their prey (insects) and with their weaving apparatus already developed;[129]

(v) Permian (299–251 MYA)—5 orders of reptiles, including turtles and tortoises; the latter appearing suddenly with its mollusk-like external skeleton;[130]

(vi) Triassic (251–200 MYA)—the order of Dinosaurs, first appearing as 6 families of Theropoda spread around the world and thus discounting a monophyletic origin, as well as the mollusk group of cuttle-fish and squids;

(vii) Jurassic (200–145 MYA)—two orders of amphibians, the Urodela (including newts) and the Anouria (including frogs); the Pterodactyls (winged reptiles); the Mesosauria (aquatic reptiles); and the first appearance of birds, namely *Archaeopteryx* and *Archaeornis*;

125. Dewar, *Illusion*, 36–37, 40–42, 45–49, 55–58.
126. Alexander, *Creation*, 125; Cooke et al., *Animals*, 28.
127. Cooke et al., *Animals*, 28; Alexander, *Creation*, 125.
128. Cooke et al., *Animals*, 28.
129. Burckhardt, *Cosmology*, 146.
130. Ibid.

(viii) Cretaceous (145–65 MYA)—16 families of bony fishes; 3 orders of teethed birds; the first appearance of placental mammals, in an order of Insectivora; and an abundance of Angiosperms, or flowering plants, dating from around 125 MYA;[131]

(ix) Eocene (56–34 MYA)—27 families of bony fishes; a large number of mammalian forms, including carnivores, odd- and even-toed ungulates, hyraxes, edentates, rodents, cetaceans, sea-cows, bats, primates, pangolins and aardvarks, as well as 11 orders that have become extinct; and 9 genera of birds without teeth.

The fossil record indicates that Angiosperms, or flowering plants in the form of around 50 families made up of hundreds of species, appear within a few million years during the middle Cretaceous. However, that the picture may be rather more complicated is suggested by the discovery of Angiosperm pollen in pre-Cambrian rocks in South America, dated to between 1.7 and 2 billion years ago.[132] Be that as it may, although Angiosperms comprise around 80% of living plant species including flowering trees, their evolutionary origin is shrouded in mystery. For instance, there are fundamental differences in the structure of both the reproductive organs and the leaves between the Angiosperms and the Gymnosperms supposedly preceding them, but there is no paleontological evidence showing any intermediates. Moreover, the two main groups of flowering plants, the monocotyledons and dicotyledons, with one or two seed-leaves, respectively, appear independently and with no known intermediates linking them.[133]

Douglas Dewar suggests that if the Darwinian hypothesis of gradual transformation was correct, the following features should be confirmed by the fossil record: (a) every class, order, family and genus would appear as a single species, exhibiting no diversity until it had been in existence for a long time; (b) the flora and fauna of any given geological period would differ but slightly from those immediately above and below in the rock strata, except on the rare

---

131. Alexander, *Creation*, 125.
132. Thompson, *Science*, 187, 191.
133. Swift, *Evolution*, 279–80.

occasions of sudden climate change, as for example inundation by the sea; (c) it should be possible to arrange fossils in various chronological series showing the origin of the animal and plant classes as well as smaller groups, thereby enabling us to accurately trace the descent of most of the species now living to the Cambrian fossils; and (d) the earliest fossils of each new group would be difficult to distinguish from those of the group from which it evolved, and the distinguishing features of the new group would be poorly developed.[134]

However, as Dewar writes, the paleontological evidence does not confirm any of these features. First, as mentioned above, new classes and orders often appear in the fossil record in great variety and not in the form of a single species. Second, although there are many cases where the fauna and flora in a particular period differ little from those in strata above and below, numerous exceptions also occur. For example, of the 3 amphibian and 9 reptilian orders living in the Permian, none survived until the Jurassic, whereas 2 and 8 new orders, respectively, had arisen by then; and of the 4 mammalian orders living in the Jurassic, none survived until the Eocene, while 20 new mammalian orders appeared in the latter epoch. Third, there is no genealogical series of fossils proving beyond doubt that any species in the past has transformed itself into a member of a different family. And, finally, the earliest known fossils of each class and order appear in the fossil record as fully developed forms, subsequent changes being relatively insignificant. Examples of this are pterodactyls, turtles, icthyosaurs, bats, cetaceans, sirenians, and seals. To the list of paleontological evidence that counters Darwinism, we might also add the significant fact that the basic body plan of vertebrates appears suddenly in the fossil record, without any recognizable predecessors.[135]

Even such a staunch Neo-Darwinist as George Simpson admits that most new species, genera and families, and nearly all the taxa above the level of families, appear suddenly in the fossil record and

---

134. Dewar, *Illusion*, 35.
135. Dewar, *Illusion*, 58–59, 61; Thompson, *Science*, 187.

Evolution According to Natural Law

without continuous transitional sequences. He mentions, for example, that all 32 orders of mammals appear fully developed in the fossil record. He added that the absence of transitional forms is a global paleontological phenomenon.[136] More recently, David Swift notes that the fossil record displays the following patterns: the abrupt appearance of distinct groups, diversification (including anomalous rates of evolution), and stasis, or the appearance and disappearance of species without their evolving into one another. However, a recognition of morphological stasis does not preclude temporal and spatial variation, which we identify with micro-evolution. Biochemically speaking, the sudden appearance of new groups corresponds with the abrupt appearance of new genetic material, which is to say that the essential protein structures were already determined by the time these groups appear on the scene.[137] This appearance of new genetic material pertains to macro-evolution, which is thus non-gradual and discontinuous in nature.

Contrary to the popular view that the fossil record provides convincing evidence for Darwinian evolution, it does not account for the appearance of a single phylum from a preceding one, since they all appear abruptly. This is also true of lower taxa such as classes and orders, while evidence of an ancestral line is found only on the levels of genera and species. However, these (micro-evolutionary) changes can be accounted for by gene segregation without the introduction of new genes. Swift therefore concludes that "although we find evidence of evolution in terms of gene shuffling, yet again we find no explanation for the origin of genetic material. The fossil record offers no evidence at all that the fundamental problem of the improbability of biological macromolecules has been overcome."[138] In other words, both paleontology and molecular biology furnish evidence against the notion that macro-evolution can occur by means of the Neo-Darwinian mechanisms of random variation and natural selection.

---

136. Smith, *Cosmos*, 70; Thompson, *Origins*, 50.
137. Swift, *Evolution*, 280, 287, 383.
138. Ibid., 295.

Biochemical Confirmation of Evolution Determined by Law

During the late twentieth century the science of biochemistry began to provide confirmation for Platonic cosmogony in the organic realms. Most biologists now accept that at least some biological forms arose spontaneously out of the self-organizing properties of their constituents, without genetic programming. Examples include the spherical form of the cell and the flat form of the cell membrane. Furthermore, the "origin of life" is an area of modern biology with a strongly deterministic element, to the extent that many researchers view biogenesis as the inevitable end of planetary evolution.[139] A leading contemporary proponent of this view is Simon Conway Morris, whose work will be considered in a later chapter.

In their 2002 paper, "The Protein Folds as Platonic Forms: New Support for the Pre-Darwinian Conception of Evolution by Natural Law," scientists Michael Denton, Craig Marshall, and Michael Legge argue that the phenomenon of protein folds provides evidence of evolution determined by law. Protein folding is "the physical process by which a protein chain acquires its native 3-dimensional structure, a conformation that is usually biologically functional, in an expeditious and reproducible manner." These structures arise when amino acids interact with each other to produce a well-defined three-dimensional structure, as determined by the amino acid sequence. Their functionality is dependent on correct folding, since misfolded proteins could lead to various degenerative diseases and also allergies.[140] According to Denton and his co-authors, the pre-Darwinian notion of organic forms provides a more powerful explanatory framework for protein folding than its selectionist successor. Protein folds are the basic building blocks of proteins, and thus of cells and all life on Earth. Each fold is a polymer consisting of 1000 to 3000 atoms folded up into a complex three-dimensional shape. In the 1970s it was discovered that protein folds might be limited in

---

139. Denton, Michael J., Marshall, Craig J. & Legge, Michael, "The Protein Folds as Platonic Forms. New Support for the pre-Darwinian Conception of Evolution by Natural Law," in *Journal of Theoretical Biology*, 219 (2002): 329–30.

140. https://en.wikipedia.org/wiki/Protein_folding.

Evolution According to Natural Law

number, while the number of three-dimensional protein structures grew significantly. The folds are therefore classified into a finite number of distinct structural families containing related but variant forms. This implies that protein folds might be natural forms determined by physical law, rather than contingent assemblages resulting from natural selection, as had previously been thought. It also became apparent that these three-dimensional structures were essentially invariant—for example, the Globin fold and the Rossman fold have remained essentially unchanged for billions of years. The facts of typology and invariance both suggest that protein folds are a finite set of timeless structures determined by physics, instead of mutable aggregates of amino acids determined by selection.[141]

The fold structures display a rational and generative morphology, with rules governing the way in which alpha helices and beta sheets can be combined into compact three-dimensional structures. Moreover, there are physical constraints restricting the folded spatial arrangements of the linear polymers of amino acids, thereby suggesting a relatively small number of permissible folds. The total number of theoretically possible protein structures formed by an amino acid chain of 150 residues long is 3^{150} or 10^{68}, while in reality the total number of stable three-dimensional structures allowed by physics is limited to around 1000 unique conformations. Therefore, Denton and his co-authors conclude, the folds represent a finite set of allowable physical structures that would recur throughout the cosmos wherever carbon-based life occurs, using the same twenty amino acids as on Earth.[142] The geneticist Richard Lewontin admits that the amino acid sequences (as coded by genes) are insufficient to explain the folding of proteins into complex three-dimensional structures. Also, while there are many alternative folded states for each sequence, only one is the physiologically active protein.[143]

There are also many cases where protein functions are clearly secondary adaptations of a primary, immutable form. It thus appears that the basic protein fold has been secondarily modified for various

---

141. Denton et al., *Protein*, 330–31.
142. Ibid., 331–32.
143. Van Vrekhem, *Evolution*, 197.

biochemical functions. For instance, the globin fold in myoglobin and various vertebrate hemoglobins entail various functional adaptations to the absorption and conveyance of oxygen. The paper contends that even the extreme Platonic view of Goethe, that form directly determines function, may be valid in cases where a particular protein function arises from the association of a particular fold with a particular prosthetic group, co-factor or metal ion.[144]

The phenomenon of protein folds also has a linguistic analogy, because the meaning of a word in a sentence depends on its context. Similarly, proteins are holistic entities, since the current evidence suggests that the various parts of the fold exert a mutual and reciprocal formative influence on each other and on the whole, which itself in turn exerts a reciprocal formative influence on all its constituent parts.[145] However, we should guard against the reductionist fallacy of identifying any organic whole only with the sum of its parts. As we are reminded by D'Arcy Wentworth Thompson, the whole is both the sum of its parts and much more: "For it is not a bundle of parts but an organization of parts, of parts in their mutual arrangement, fitting one with another, in what Aristotle calls 'a single and indivisible principle of unity'; and this is no merely metaphysical conception." Thompson opposes this "fundamental truth" to Darwin's view that natural selection is capable of both reducing or developing any part of an organism without a corresponding and compensatory development or reduction.[146] We should add that the interaction between an organic whole and its parts could be viewed in metaphysical terms as a biochemical manifestation of the interaction between the One and the many that constitutes the cosmos.

The process of protein folding appears as matter drawn into a pre-existing Platonic mold: "During folding the amino acid sequence of a protein appears to be searching conformation space for increasingly stable intermediates which lead it step wise toward the deepest energy minimum for that sequence, which corresponds

---

144. Denton et al., *Protein*, 332.
145. Ibid., 335.
146. Thompson, *Growth*, 264.

Evolution According to Natural Law

to its final native conformation."[147] This is analogous to a ball finding its way down the sides of an irregularly-shaped bowl to the bottom, with the bottom representing the free energy minimum of the fold. The standard claim that the amino acid sequence determines the three-dimensional form of a protein is a mechanistic interpretation of the folding process, but Denton and his co-authors claim that the folding process is more accurately depicted in Platonic terms, in which the prior laws of form determine which amino acid sequences can fold into a stable three-dimensional form.[148] As we noted in an earlier chapter, Aristotle similarly employs the analogy of a pre-existing plan of a house that makes use of the building materials in conformity with the plan. Thus, although the bricks and stone are chronologically prior to the house, logically the form of the house comes first.

It has been observed that the three-dimensional conformations of protein folds are quite resistant to evolutionary changes in their amino acid sequences. Evidently the laws of physics allow only a limited number of folds, yet numerous apparently unrelated amino acid sequences can fold into the same form. Denton and his fellow authors contend that this self-organization into the same fold by very different amino acid sequences confirms the Platonic primacy of the protein fold over its material constituents. Moreover, the robustness of protein folds holds definite evolutionary implications, this being a natural, intrinsic feature of the folds and not a secondarily evolved feature. In this way the folds provide the evolutionary process with stable structures upon which to build more complex structures and functions.[149]

In view what we have said above, it might be thought that the 1000-odd protein folds represent a physically determined bottleneck through which protein evolution had to pass. This would also be true on any Earth-like planet where proteins are constructed out of the same twenty amino acids. Furthermore,

---

147. Ptitsyn & Finkelstein, 1980; Finkelstein & Ptitsyn, 1987; quoted in Denton et al., *Protein*, 332.
148. Denton et al., *Protein*, 332.
149. Ibid., 334.

If it is possible to derive folds via evolutionary constructional sequences starting from say a simple alpha helix and leading via a double helical motif and so on, then selection for biological functions may have played at least some role. However as many authors have pointed out, in the context of protein evolution, selection must have a detectable proto-function to start with. This means that before selection begins there must be at least some sort of stable scaffold on which a function can be hung.[150]

Moreover, Dean Kenyon and Gary Steinman argue in their work *Biochemical Predestination* (1969) that the ultimate development of a living cell is determined by the physical and chemical properties possessed by the starting compounds from which these systems evolved,[151] which is as much as to say that cells develop according to natural law and not by means of random processes.

In summary, the protein folds represent a finite set of around 1000 natural forms determined by the laws of physics, like atoms and crystals. Therefore, they do not conform to the Darwinian notion of organic forms as contingent, functionally contrived assemblages of matter. The protein folds thus represent the first case in the history of biology where a set of complex organic forms can be shown to be unambiguously lawful natural forms in the traditional, pre-Darwinian sense. It thereby challenges the Darwinian dictum that all complex organic forms are contingent, artefact-like products of selection.[152]

Further biochemical evidence in favor of regulated evolution is found in cell forms. For example, the cell form of the genus *Tetrahymena* (free-living ciliate protozoa which are common in freshwater ponds)[153] has remained basically unchanged for around a billion years, despite the fact that the molecular constituents of its species vary enormously. We thus observe an invariance of cell form with a marked variation in its building blocks, which suggests that certain features of the *Tetrahymena* cell form might be non-functional

150. Ibid., 337.
151. Ibid., 338.
152. Ibid., 338–39.
153. https://en.wikipedia.org/wiki/Tetrahymena.

structures determined by law, formed by physics rather than selection. Another example is the extraordinary ability of cells like *Stentor* (a genus of ciliate protists that can reach a length of two millimeters, making them among the largest extant unicellular organisms)[154] to recover their "proper form" after micro-surgical manipulations. It is therefore possible that the whole cell is behaving like a natural form, searching conformational space.[155]

In the universe of protein folds it thus appears that functional adaptations are secondary modifications of the essentially invariant data of physics; their evolution takes place by law, not by selection for a function. "It is a universe where abstract rules, like the rules of grammar, define a set of unique immaterial templates which are materialized into a thousand or so natural forms—a world of rational morphology and pre-ordained evolutionary paths."[156] As Denton and his co-authors conclude, it thus represents a pre-Darwinian Platonic universe.

---

154. https://en.wikipedia.org/wiki/Stentor_protozoa.
155. Denton et al., *Protein*, 340.
156. Ibid.

7

Directed Evolution

SINCE THE LATE nineteenth century, one of the most important scientific alternatives to Darwinism has been orthogenesis, or directed evolution. The term orthogenesis was coined by the German zoologist Wilhelm Haacke in his 1893 work *Gestaltung und Vererbung*, and is derived from the Greek *orthos*, meaning straight, as "in a straight line." Its primary meaning correlates with the modern phrase "constraints on variation." In his work, Haacke presents a theory of heredity similar to the "germ plasm" theory of the biologist August Weismann, but holding that genetic material combines in such a way that organisms can only vary in definite directions. It is this capacity that Haacke terms *orthogenesis*, contrasting it with *amphigenesis*, or the ability to vary in every possible direction.[1] Orthogenesis was popularized by another German zoologist, Theodor Eimer, a former student of Weismann who later came to reject his mentor's adherence to Darwinism. After an extensive study of butterflies, Eimer published *Orthogenesis der Schmetterlinge* (1897), in which he argues that evolution occurs almost exclusively as development along definitely determined lines.[2]

The theory of orthogenesis postulates that "living organisms have a predisposition to vary in certain directions, and this very predisposition determines the trend of evolution, first of all, irrespectively from adaptation and selection; as the crystals grow taking a certain form, so phylogenetic trends evolve following their internal laws,"

---

1. LSJ, 497; Popov, *Constraints*, 205–06.
2. Bergman, Jerry, *The Rise and Fall of the Orthogenesis Non-Darwinian Theory of Evolution* (2009): 141. https://www.creationresearch.org/crsq/articles/47/47_2/CRSQ%20Fall%202010%20Bergman.pdf.

writes the Russian scientist Igor Popov.[3] It was actually recognized by an opponent of orthogenesis, Theodosius Dobzhansky, that "if evolution is orthogenesis, then it is what the etymology of the word 'evolution' implies, i.e., unfolding of pre-existing rudiments, like the development of a flower from a bud." The famous Neo-Darwinist also mentions the evidence of progress and directionality found in the living world as a whole. Exceptions to this are various parasites of which the evolution is retrogressive, as for example in the degeneration of the nervous system, or groups in which evolution has produced endless variations on the same theme. Yet the net outcome of evolution is that the Earth is no longer populated by single-celled or simple multicellular organisms only. Instead, Dobzhansky adds, we observe numerous complex organisms with body structures that are comparable to works of art, as well as organisms with highly developed nervous systems, which enable them to dominate their environment to some extent.[4]

However, faced with the evidence of retrogression just mentioned, Dobzhansky then states that the harmfulness of most mutations demonstrates the absence of guidance in evolution. He says: "At the level of mutation, evolution is neither directional nor oriented nor progressive. It is the very antithesis of orthogenesis. Mutation alone would cause chaos, not evolution. Natural selection redresses the balance."[5] In classical Darwinian fashion, natural selection is thus called upon to redress the alleged deficiencies of other evolutionary mechanisms, such as mutation. More recently some Neo-Darwinists have tried hard to find a selectionist explanation at any price, so that even the finest details of the vein structure in the wings of fruit flies are attributed to natural selection.[6]

Nonetheless, it cannot be denied that adaptation plays a significant role in micro-evolution. The celebrated case of the Galapagos finches is an example of undirected yet adaptive biological change,

---

3. Popov, *Constraints*, 205.
4. Dobzhansky, *Biology*, 117–19.
5. Ibid., 122.
6. Popov, *Constraints*, 213.

as Michael Denton has remarked. Some of these species comprise the genus *Geospiza*, and genetic studies have confirmed Darwin's hypothesis that all are descended from a single species that arrived from South America. As the dominant birds on these islands, the finches have diversified to occupy many ecological niches, and their small size and the distances involved mean that there is limited breeding between the various island populations. However, it has been established that the finch species regularly produce hybrid offspring, which implies that the differences in morphology and behavior are attributable to gene segregation rather than mutation. The Galapagos finches should thus be recognized as a case of limited or micro-evolution in which natural selection had a role to play in their diversification from a common ancestor.[7]

Development of the Theory

During the nineteenth century the arguments in favor of directed evolution were based mainly on the observation of a restricted number of directions in variation, such as the existence of non-adaptive characters, the phylogenetic regularities as evinced by the fossil record, and the phenomenon of parallel evolution.[8] The neglected Alfred Wallace, who shared the formulation of the theory of natural selection with Charles Darwin, eventually became a proponent of directed evolution. He believed that the complexity of the cell cannot be explained in terms of matter and mechanism only, but is due to a directed cause. Noting that the fertilized cell nucleus is the seat of heredity and development, the English naturalist posed the following pertinent questions regarding causality: What is the agency that sets in motion a whole series of mechanical, chemical, and vital forces, and guides them at every step to their destined end? What power gave life to the living protoplasm out of which the cell consists, and organized the highly differentiated nucleus? In cell division, what power determines the early cell-mass to assume well-defined shapes? Who or what guides the atoms of protoplasmic

---

7. Denton, *Destiny*, 287; Cooke et al., *Animals*, 353; Swift, *Evolution*, 227, 377.
8. Popov, *Constraints*, 205.

Directed Evolution

molecules into new combinations chemically and new structures mechanically, e.g., muscle and bone?[9] Wallace then reasons,

> But this orderly process is quite unintelligible without some *directive organizing* power constantly at work in or upon every chemical atom or physical molecule of the whole structure, as one after another they are brought to their places, and built in, as it were, to the structure of every tissue of every organ as it takes form and substance in the fabric of the living, moving, and, in the case of animals, sensitive creation.[10]

Wallace's notion of a directive organizing power indwelling every atom and cell recalls, albeit in scientific terminology, the Hellenic notion of the reason-principles (*logoi*) that indwell all living things and provide intelligibility to the organic realms.

In the early twentieth century the theory of orthogenesis began to receive increasing support from the results of experimental breeding with various plant and animal species. A notable role was played in this regard by the American zoologist Charles Whitman, who studied the variation of pigeons extensively and concluded that the variability of pigeons displays various regularities confirming the reality of orthogenesis. For instance, a reduction of pigmentation in pigeon plumage was obtained step by step over several generations in order to produce white pigeons. Whitman wrote, "Such a reduction always takes place in a definite direction: from the forepart of the body through the spotty variant and the variant with two stripes on the wings. The pigment was not lost evenly and gradually in the whole surface of the pigeon, and it did not disappear in any other direction."[11] If the Darwinians were correct with their insistence on infinite variability, then the pigeon-breeding would not have produced such a directed change in coloration. Whitman's contemporary, the eminent paleontologist Henry Osborn, also advocated orthogenesis, basing his arguments on the evidence of parallel evolution, also known as convergent evolution. Convergence refers to

---

9. Flannery, *Wallace*, 28, 147–48.
10. Quoted in Flannery, *Wallace*, 148.
11. Quoted in Popov, *Constraints*, 206.

the independent development of similar structures in unrelated groups of animals. For example, wings have evolved independently at least three times, among pterodactyls, birds, and bats. Instances such as these convinced Osborn of the existence of an orthogenetic inner drive that produced remarkably similar structures even in widely different environments. He viewed evolution as an explosion out from an ancestral form, such as occurs in adaptive radiation. Moreover, once a structure has reached a certain level of perfection its evolution comes to an end and therefore, Osborn argues, it is impossible for an organism to adapt to a different physical environment.[12] This would explain the existence of so-called living fossils, which have undergone little change since their first appearance in the fossil record. Perhaps the most striking instance is the reef-building bacteria known as stromalites or blue-green algae, which first appeared in the fossil record around 3.5 billion years ago. Stromatolites were formed mainly by photosynthesizing colonial cyanobacteria and were the dominant life-form on Earth for an estimated two billion years. Astonishingly, living stromalites are still found today, for example at Shark Bay in Western Australia.[13]

The Russian contemporary of these American scientists, Lev Berg, asserts that the variation of characters in an evolutionary lineage follows a definite course, like an electric current moving along a wire. He illustrates this with examples from both the plant and animal kingdoms. In the plant kingdom a gradual reduction of sexual (gametophyte) generation and a corresponding increase of asexual (sporophyte) generation has occurred. This is evident from the well-developed prothallus of terrestrial ferns and horse-tails, to the reduced prothallus in heterosporous ferns and club-mosses, to the sporophyte of gymnosperms. A definite course of evolution is thus displayed in the reduction of the gametophyte—from a flourishing condition in mosses, to a gradual decline in ferns, to a complete disappearance in gymnosperms, to its final replacement by the sporophyte in angiosperms. Significantly, these are not links of one

---

12. Bergman, *Orthogenesis*, 141–42.
13. Cooke et al., *Animals*, 22.

Directed Evolution

genealogical chain—ferns, for example, are older than mosses—but various genetic branches.[14]

Turning to the animal kingdom, Berg finds evidence for directed evolution in many evolutionary trends among vertebrates. Examples include the evolution of teeth among reptiles and mammals; the gradual ossification of the vertebral column, for example in fishes; the reduction in the number of bones in the skull; the transformation of a two-chambered heart into a three- and four-chambered organ, with a corresponding increase in the complexity of the circulatory system; and the evolution of the brain. Among vertebrates the skull, vertebral column, and extremities developed in a definite direction, without the presence of chance variations being selected by nature. Also, in the evolution of arterial and venous systems in vertebrates from lampreys and fishes to mammals, we find that the number of pairs of aortic arches gradually diminishes. Finally, the resemblance between crocodiles and birds in respect of the heart is also manifested in the structure of the brain.[15] As French zoologist Pierre Grassé remarks, "The existence of oriented lines is a fact, and not a theoretical view; a line can only be identified and exists solely because it embodies a given trend appearing in individuals which derive from one another and succeed one another in time."[16]

Berg describes the course of evolution as follows: "A given group of organisms in the course of time breaks up into forms which either repeat the course of development of the existing forms, or follow a direction in which will subsequently develop still more highly organized groups," which is to say that there are a restricted number of definite possibilities, while allowing for an immense variety in details. For Berg this is due to the fact that internal, i.e., biochemical, causes predominate over the environment, for example in leading an organism to destruction or in the entire process of embryonic development. Also, in sea dwellers such as ammonites, the same mutations are produced in both stable and changing envi-

---

14. Berg, *Nomogenesis*, 110, 118–21.
15. Ibid., 121, 124–27.
16. Quoted in Davison, *Manifesto*, 33

ronments. The insignificant influence of climate on organic evolution is demonstrated by the glacial epoch, which had a catastrophic effect on the geographical distribution of species through extinction or migration, but in which nature produced very few new forms.[17] We thus conclude that the observed direction in evolution is primarily due to internal factors, and only secondarily to environmental factors.

According to Henry Osborn, there is no paleontological evidence that the fit originates from the fortuitous by natural selection. Berg, too, argues against natural selection as the tool of evolution using the example of the deer (genus *Cervus*), which arose from different species, or polyphyletically, in the Miocene. Since polyphyletic evolution excludes the operation of chance in the production of characters, it is irreconcilable with the principle of selection. Therefore, "every polyphyletic genus affords the most obvious evidence in favor of the assumption that development follows a definite course, that it could not proceed in any other way than the one it has taken, that variations are not infinite, but strictly limited in number, and that polyphyletic evolution is not the exception, but the rule."[18] This is diametrically opposed to the Darwinian stance on monophyletic evolution, as noted earlier. After surveying numerous examples from the plant, animal, and human kingdoms, Berg concludes that (i) new characters arise not accidentally, but in accordance with law, (ii) the struggle for existence and natural selection had nothing to do with the development of these characters, and (iii) evolution proceeds in accordance with law, i.e., in a determined direction.[19]

One of Berg's Russian colleagues, the botanist Nikolay Vavilov, undertook an extensive study of cultivated plants and discovered that in different species of cereals, parallel variations occur in the shape of the ears and the color of the seeds. This phenomenon even occurs among unrelated plants, such as identical variations of the root form in beet, carrot and turnip. Furthermore, variations such as gigantism, dwarfism, fasciation, and albinism are found through-

---

17. Berg, *Nomogenesis*, 114–16, 163.
18. Ibid., 127–28.
19. Ibid., 107–08.

out the plant kingdom. This phenomenon of parallel variations even enabled the Russian botanist to predict the discovery of new plant forms: since lobate forms were known among pumpkins and melons, but unknown among watermelons, Vavilov predicted the discovery of lobate watermelons, which were duly found in southeastern Russia.[20] Evidently the theory of directed evolution harbors predictive power, thereby meeting a crucial requirement for a valid scientific theory.

The phenomenon of parallel variability was known to Darwin, who ascribed it to a common origin of the species in question. Vavilov agreed with Darwin on the common origin of homological series, but argued that there are defined directions in variability and therefore in evolution. "Moreover," writes Popov, "he [Vavilov] paid attention to the fact that different mutations could create identical phenotypes, i.e., the number of possible phenotypes is smaller than the number of possible mutations." It is not surprising that the eminent Russian proponents of directed evolution such as Berg, Vavilov and the botanist Sobolev collaborated with each other. Unfortunately their studies were limited by political conditions, as Popov noted, since Darwinism occupied the same place as Communism in Soviet orthodoxy. Nonetheless, from the 1970s there was a revival of interest in Berg's work among Soviet scientists, which led to the publication of some of his further writings on evolutionary theory.[21]

In Germany the geneticist Viktor Jollos was led by his research data to argue in favor of orthogenetic mutations. He published an essay in 1931 showing that the wild type of the fruit-fly *Drosophila* produced a mutation of eye color in the sequence dark → light → yellow, ivory and white. No other sequences were observed, Popov writes, so that Jollos termed this phenomenon "directed mutation." Tragically, Jollos was forced to emigrate from National Socialist Germany in 1933 and died a few years later in America, Vavilov died in a Soviet prison, and Whitman died before his studies on pigeons

---

20. Popov, *Constraints*, 206.
21. Ibid., 207, 210–11.

could be completed.[22] The untimely deaths of these orthogenetic scientists thus contributed, albeit indirectly, to the rise to dominance of the Neo-Darwinian synthesis with its emphasis on random variations.[23]

Remarkably, the belief that selection could create practically anything led some of the scientists who played leading roles in the formation of the Neo-Darwinian synthesis to downplay indications of evolutionary constraints. This is evident, for instance, in the cases of Theodosius Dobzhansky and Ernst Mayr. In the first editions of their influential works *Genetics and the Origin of Species* (1937) and *Systematics and the Origin of Species* (1942), respectively, these biologists devoted entire chapters to the question of evolutionary constraints. However, in further editions and writings this question is reduced and finally disappears altogether. Thus did the Neo-Darwinists comply with Darwin's dictum in *The Variation of Animals and Plants under Domestication*, 1883, that variability is subordinate to selection.[24]

The notion of direction in the evolutionary process later found support from Arthur Koestler in his *Ghost in the Machine* (1967). The Hungarian-born author cites scientists such as Ludwig von Bertalanffy, a founder of general systems theory, who viewed evolution not as a process based on selection within a chaotic mass of mutations (as Darwinism holds), but as one governed by definite laws; and Helen Spurway who argued that the evidence from homology suggests a restricted mutational spectrum which determines the possibilities of evolution. Accordingly, given our planetary conditions, atmosphere, energies and building material, life from its first inception could only progress in a limited number of ways. Koestler concluded that the evolution of life is a game played according to fixed rules, which are inherent in the structure of living matter.[25]

---

22. Berg was an exception, having been an honored member of the Soviet academic establishment due to his monumental work on the fishes of the USSR.
23. Popov, *Constraints*, 207–08.
24. Ibid., 208–10.
25. Denton, *Destiny*, 272.

Directed Evolution

Constraints on Variation

The theory of directed evolution is closely linked to the notion of constraints on variation. In his paper *The Problem of Constraints on Variation*, Igor Popov notes that the Darwinian scheme of mutation-selection contains a gap, namely the distance between the mutation and the phenotype exposed to selection. This was recognized by some Darwinists, who then employed developmental biology to fill the gap, since this discipline exposed the special forces influencing the path from gene to phenotype. One means of filling the gap is to posit constraints on the number of possible directions of variation.[26]

Darwin held that the variability of organisms is infinite, or, as Swift says, for Darwin biological tissues possess an innate "plasticity" which is fundamentally unlimited. Lev Berg objects that this notion of infinite variability is refuted by the paleontological evidence. For instance, the fossil remains of the molar teeth of mammals indicate that the majority of teeth are derived from a primitive type of tritubercular teeth, in orders as varied as Insectivora, Ungulata, and Primata. Thus the complexity of the molars develops in strict accordance with a law, with the supplementary tubercules and their variations appearing in a determined order and a definite position. Furthermore, among ammonites a limited number of variations are arranged in definite lines. Thus in the species *Phylloceras heterophyllum* we observe an ever-growing complexity of the lobate line from more ancient to more recent deposits. However, according to Swift, in Darwin's time it was not known that the apparent "plasticity" of tissues results from underlying genetic and biochemical mechanisms. We now know that morphological variations arise through different combinations of genes. This finite pool of available genetic material thus limits the extent of possible variations.[27]

According to Stephen Jay Gould, the science of developmental biology demonstrates the conservation of basic pathways of development among phyla that have been evolving independently for at least 500 million years while being very different in basic anatomy,

---

26. Popov, *Constraints*, 213.
27. Swift, *Evolution*, 299, 314; Berg, *Nomogenesis*, 25–26, 385.

From Logos to Bios

such as insects and vertebrates. Some examples of this conservation of pathways are as follows: (i) The homeotic genes of fruit flies are also present in vertebrates, functioning in effectively the same way; thus, their mutations disturb the order of the parts along the main body axis, as for example the legs. (ii) The same gene conserves and mediates the major developmental pathways for eyes in animals as diverse as squids, flies and vertebrates, although the final products differ markedly. (iii) The same genes regulate the formation of top and bottom surfaces in insects and vertebrates, although with an inverted order, so in vertebrates the spinal cord is placed above the gut, whereas in insects the main nerve cord is placed below the gut. Gould concludes that this genetic stability acts as a constraint upon the range and potentiality of adaptation, with such a constraint being at least of equal importance to the advantages of adaptation.[28]

Neither Darwin nor the Neo-Darwinists were able to solve the problem of constraints on variation, as Popov remarks. Instead, they strove to conceal or discredit any data related to constraints. They had good reason for doing so, for "otherwise, if recognized, the limitation to variation will always and everywhere affect evolution, and this means that evolution is a movement on rails, instead of wandering through the vast space of adaptation. If that is the case, selection should be considered a destructive force. Moreover, such a standpoint is not Darwinism anymore, it is orthogenesis, and it cannot be included into the Darwinian paradigm."[29] This image of evolution as a movement on rails is quite a useful one, since it also accommodates the phenomena of evolutionary dead-ends (= a railway terminus) and extinctions (= an accidental derailing).

Against the Darwinian notion of infinite variability, Lev Berg emphasized that nature can only produce new forms from genotypical, not phenotypical, material. Mutations, or inherited variations, are required. Furthermore, the observed number of mutations is so limited that there is no scope for selection to operate. Berg adds that infinite variations would have led to deformations and monstrosities as the norm, and well-adapted beings the exception. He writes,

---

28. Gould, *Fundamentalism*, 4–5.
29. Popov, *Constraints*, 216.

Directed Evolution

"An organism is a stable system, in which a tendency towards variation is confined within certain limits by inheritance.... It would be impossible to conceive how such complex organs as the eye, the ear or the pituitary body could properly exercise their functions, if they were the seat of an infinite number of variations, from which it would be left to chance to select the most efficient." On all the steps of the phylogenetic ladder are found beings perfectly adapted to their environment, and not the monstrous forms which would have followed from infinite variability. The extraordinary polymorphism displayed by wheat, rye, barley and lentil is due to the redistribution of the same characters, while the boundary-lines separating the different species remain inviolable. This intra-specific variability is determined by natural law, so that similar forms are found among different species.[30]

That constraints on variation similarly apply to artificial selection has been demonstrated by the horticulturist Luther Burbank in his plant-breeding experiments which led to the discovery of a "pull" toward the mean that keeps all living things within some more or less fixed limitations. He called this the Law of the Reversion to the Average. This phenomenon was also encountered by Ernst Mayr in his experiments with fruit flies. He managed to markedly increase and decrease the number of bristles on their bodies, but without selective breeding, the carriers of these extremes either died out or reverted to their average. As Richard Thompson writes, "These results reveal a major anti-evolutionary characteristic of species: when changes are pushed beyond a certain limit members of a species will become sterile and die out or else revert to their standard form."[31] The Law of Reversion is also demonstrated by the role natural selection plays when domesticated animals are returned to the wild. The deviant selected forms rapidly disappear in favor of the more standard types which more closely resemble their ancestors.[32]

The notion of constraints on variation received significant buttressing from D'Arcy Wentworth Thompson, whose epochal work

---

30. Berg, *Nomogenesis*, 27, 29–30.
31. Thompson, *Origins*, 45–46.
32. Davison, *Manifesto*, 12, 33.

On Growth and Form we discussed in an earlier chapter. In it the Scottish polymath depicts an immense variety of organic forms in terms of mathematical patterns and physical laws. Igor Popov notes that this approach enabled Thompson to describe in considerable detail the possible limitations to organic forms, determined by their physical conditions. Although in this work Thompson does not engage in anti-Darwinian polemics, he does discuss various points of disagreement with Darwinism. For example, given his holistic view of organic reality Thompson rejects the notion of independent single characters, arguing instead that all the parts of an organism are correlated with the others. He also denies the adaptationist explanations of organic forms such as the sponge-spicule and Radiolaria, asserting that they are due rather to chemical and physical characteristics. And regarding Radiolaria, since several kinds are carried along by waves, "any explanation of the evolution of their form in terms of the Darwinian struggle for existence becomes meaningless."[33]

An apparently universal feature of orthogenesis is that new life forms typically appear as small organisms that subsequently become larger and more specialized, as John Davison observes. Examples of this tendency are found among dinosaurs, titanotheres and ammonites. This had been discussed by the paleontologist Otto Schindewolf, who identified three phases in the evolutionary process. The first, *typogenesis*, involves the rapid establishment of new forms. The slower, second phase, *typostasis*, entails elaboration and diversification of the newly-established forms. The third phase, *typolysis*, is characterized by gigantism and over-specialization, and, not surprisingly, ends with extinction.[34] Schindewolf's scheme serves to explain various trends in the fossil record that appear to be incompatible with natural selection and adaptation. Many cases of extinction have been attributed to the evolution of characters that were excessive for their possessors. For example, the massive reptiles of the Cretaceous era required enormous quantities of food merely to survive; the antlers of the extinct Irish elk (genus *Megaloceros*)

---

33. Popov, *Constraints*, 208.
34. Davison, *Manifesto*, 37–38.

were so large that the animal had difficulty walking; and the nasal horns of the rhinoceros-like titanothere evolved to such an extent that they seriously interfered with adaptation.[35]

David Swift posits a genetic basis for over-specialization, claiming that this phenomenon is not due to the gain of genetic material, but to its loss. He argues that in some cases evolutionary success has been due to the intensive selection of a very limited set of genes, which increases susceptibility to extinction when circumstances change. A prime example of this today is the cheetah (*Acinonyx jubatus*), the fastest land mammal on Earth. With its long legs and non-retractile claws, this big cat is well-adapted to living in open grassland in sub-Saharan Africa, where its speed enables it to run down the smaller antelopes that comprise the bulk of its prey. The cheetah's reputation as the most specialized living cat is probably due to its low genetic diversity (the lowest among the big cats), for which natural selection had to compensate by dispensing with unnecessary features, such as the retractile claws of other cats.[36] However, due to the destruction of its specialized habitat by human activities, as well as fierce competition from lions and hyenas, the cheetah's conservation status has become vulnerable. In contrast, the leopard (*Panthera pardus*) is the most widespread big cat, living in a variety of habitats from tropical jungle to semi-desert to mountains across Africa and southern Asia, and feeding opportunistically on a wide range of prey. As a result, the conservation status of the leopard is locally common, being widespread and abundant within its range.[37] Unlike the leopard, the cheetah is unlikely to survive outside protected areas in the long run.

In support of orthogenesis, Lev Berg noted the gradual reduction of the hind limbs and pelvis in aquatic mammals of the orders Cetacea (whales) and Sirenia (dugong and manatees). Also, in the evolutionary series of the horse from *Orohippus* of the Eocene epoch to the present-day genus *Equus*, the dentition displays a direction in that the widest tooth moves from the molars to the premolars. The

---

35. Bergman, *Orthogenesis*, 141.
36. Swift, *Evolution*, 253–54, 294.
37. Cooke et al., *Animals*, 153, 155.

fossil record also shows the directed development of the Nautiloid cephalopods from forms with a straight shell to those with a spiral shell.[38] Since the 1920s, scientists have often cited the shells of various invertebrate groups as examples of directed evolution, beginning with the Russian entomologist Yuri Filipchenko, who, incidentally, also coined the terms micro- and macro-evolution. Noting that the shells of foraminifera are similar enough to those of extinct cephalopods that they were initially considered to be mollusks, Filipchenko came to the conclusion that "the laws of growth and step-by-step enrollment of primary straight shells are probably equal," whether "the shell belongs either to a unicellular rhizopod or to a highly developed representative of mollusks."[39] This provides further evidence that evolutionary constraints are based upon organic law.

The evidence from shells was likewise employed by Stephen Jay Gould and Richard Lewontin in a 1979 essay impressively titled *The spandrels of San Marco and the Panglossian paradigm: a critique of the adaptationist program*. The "Panglossian paradigm" refers to the adaptationist program that was popularized towards the end of the nineteenth century by Alfred Wallace and August Weismann. As Gould and Lewontin write, "This program regards natural selection as so powerful and the constraints upon it so few that direct production of adaptation through its operation becomes the primary cause of nearly all organic form, function, and behavior." The essay also considers the work of the German paleontologist Adolf Seilacher and his notion of "architectural constraints," or morphological restrictions that were never adaptations. As an example, the divaricate (branching) form of architecture occurs repeatedly in the shells of all groups of mollusks, as well as in the unrelated brachiopods. It manifests in a variety of structures, including raised ornamental lines, patterns of coloration, and incised grooves. Seilacher contended that most manifestations of this pattern are the result of inhomogeneity in the growing mantle of the organism and are therefore non-adaptive. This is suggested by, for example, the color

---

38. Berg, *Nomogenesis*, 128–30.
39. Quoted in Popov, *Constraints*, 212.

Directed Evolution

patterns of clams that are invisible due to the animal living buried in sediments or remaining covered with a thick layer of periostracum. Their divaricate structure is therefore a fundamental architectural constraint, which could be useful to the organism but cannot be the result of adaptation.[40]

Michael Denton and Perry Marshall both observe that a further constraint against undirected change is the phenomenon of genetic redundancy. This might be understood as a genetic backup system, where organisms are reinforced by partially or totally redundant genes to guard against random mutational error. Redundancy is built into the very structure of DNA, Marshall writes. Since genetic data is stored in the four "letters" of adenine (A), cytosine (C), guanine (G) and thymine (T), and these are arranged in groups of three called codons or triplets, it follows that there are 64 ($= 4^3$) possible triplets, or "words" in the genetic vocabulary. And since these 64 letters provide the instructions to build 20 different amino acids, it implies that DNA has a 3:1 back-up system to guard against the errors that would probably result from single-letter copying. Furthermore, organs such as the female vulva develop by means of two or more quite different mechanisms, of which any one is sufficient to generate a perfect organ. Denton notes that this implies that organs cannot be radically transformed through a succession of small, independent changes as postulated in Darwinism, since simultaneous changes are required in various mechanisms. The same requirement applies to protein molecules, in which each atom co-operatively interacts with all other atoms, so that the proteins form complex holistic systems.[41]

As a matter of fact, it has recently become evident in molecular biology that living systems are best understood in holistic or network terms, in which the properties of the networks are not merely the sum of the individual parts. Instead, the properties are emergent, so that the features of each network are only intelligible in a holistic way. For instance, in genomic databases it is not the genes alone, or the proteins encoded by the genes, but the interactions

---

40. Gould, *Life*, 419, 432–34.
41. Marshall, *Evolution*, 52–54; Denton, *Destiny*, 337–40.

between the proteins and various other macro-molecular components in each cell that constitutes the network.[42]

According to a 2001 study by Lev Yampolsky and Arlin Stoltzfus, the existing models of population genetics (a key component of the Neo-Darwinian synthesis) are biased since they assume *a priori* that variations will be available and thus underestimate mutation pressure. In reality, evolution is driven by mutation, which produces novelty. The authors concluded that the phenomena of homoplasy, parallelism and directionality are due to internal factors—just as Lev Berg had insisted 80 years earlier. In another study published in 2003, Elena Kovalenko collected a quantity of empirical material which did not correspond with Darwin's view on the properties of variability. The Russian molecular biologist found that the number of available variations is significantly less than the number that could be calculated theoretically. For example, mass breeding and field-work revealed that various species of frog (genus *Anura*) showed 45 variations out of the possible 288. Moreover, for each species not only the number and frequency but also the qualitative structure of variations were constant in samples from different populations, in their offspring, and in both young and adult frogs. Kovalenko concluded that natural selection could play a limited role as an exception, with the species still represented by those variations which have a higher probability in the spectrum of its variability.[43] Once again, we reach the conclusion that natural selection does contribute to evolution, but in a role secondary to mutation.

One of the most important thinkers on evolutionary theory today is the already-mentioned paleontologist Simon Conway Morris, who achieved renown with his work on the Burgess Shale fossils. This Canadian site contains the most extensive fossil deposits of the Cambrian era, with more than 60,000 fossils having been recovered since its discovery in 1909.[44] In his book *The Crucible of Creation* (1998), Conway Morris argues that these early animals became extinct because they were ill-adapted to their environment. This lat-

42. Hewlett, *Biology*, 182.
43. Popov, *Constraints*, 212, 215–16.
44. Cooke et al., *Animals*, 27.

Directed Evolution

ter is always the result of physical and biochemical laws, which constrain the types of organisms to develop within certain limits. These laws will therefore ensure that the carbon atoms found in stars will eventually combine into long molecules capable of replication, that these molecules will build bodies sensitive to heat and light, and that these bodies will be capable of movement in their environment, with their sense organs situated towards the front end.[45] In other words, the physical environment as such acts as a (ubiquitous) constraint on variations.

Along similar lines, Michael Denton and Craig Marshall argue in a 2001 essay that the evolution of biological forms has to a large extent been determined by physical law. This implies that the immense diversity of life-forms is underpinned by a finite set of natural forms that will recur anywhere in the cosmos where carbon-based life exists. Conway Morris concurs:

> Not all is possible, options are limited, and different starting points converge repeatedly on the same destinations.... The "landscape" of biological form, be it at the level of proteins, organisms, or social systems, may in principle be almost infinitely rich, but in reality the number of "roads" through it may be much more restricted.[46]

Biochemical Evidence

More recent evidence in favor of directed evolution includes the uniformity and speed of molecular evolution, and the rapidity of the major morphological transitions. Apparently unaware of this, the biologist Jerry Bergman asserted that the theory of orthogenesis has been largely abandoned, since no known mechanism exists to account for an endogenous perfecting force.[47] In fact, Michael Denton contends, the evidence from genetics suggests that the whole pattern of evolution might have been written into the DNA of all fauna and flora from the outset. Thus every living organism is spec-

---

45. Ward, Keith, *The Big Questions in Science and Religion* (West Conshohocken, PA: Templeton Foundation Press, 2008), 66.
46. Conway Morris, *Life*, 11.
47. Denton, *Destiny*, 265; Bergman, *Orthogenesis*, 144.

ified in a precisely-determined way through a set of instructions encoded in the sequence of bases in its DNA. The available evidence implies that DNA is fit not only for heredity but also for directed evolution. Also, as Weismann has shown, it is suggestive that biological information flows in only one direction, namely from genes to organism, or, more precisely, information flows from DNA to protein, that is to say from linear DNA space to three-dimensional morphological space. As Denton writes, "This unidirectional flow of information from DNA to organism is clearly 'fit' for directed evolution."[48] This calls to mind the Pythagorean and Platonic precept that the cosmos arises through a progression from two-dimensional geometrical surfaces to three-dimensional solid bodies.

Neo-Darwinian orthodoxy, on the contrary, insists on the spontaneity of mutation, which is viewed as comprising accidental, undirected and random events, and is moreover non-oriented with respect to adaptation. The only evidence for "spontaneous" mutation is provided by the Luria-Delbrück experiment conducted in 1943, which found that in the bacterium *Escherichia coli* genetic mutations arise in the absence of selection. This was interpreted by Neo-Darwinists as confirmation of the theory that natural selection acting on random mutations applies not only to more complex organisms, but to bacteria as well. However, Max Delbrück later admitted that specifically adaptive mutations are possible. Whichever way, Denton remarks, it is impossible to demonstrate experimentally that all of the changes in DNA sequence over the past four billion years of evolution have occurred spontaneously, although some undoubtedly did.[49] Regarding *E. coli*, a lengthy study from 1988 until 2014 of 12 bacterial populations revealed that some adaptations occurred in all of them. This evidence suggests, Perry Marshall writes, that identical adaptations took place multiple times.[50] This provides further evidence in favor of directed evolution.

It is also noteworthy that unlike evolutionary change at the mor-

---

48. Denton, *Destiny*, 275–76.
49. https://en.wikipedia.org/wiki/Luria-Delbrück_experiment; Denton, *Destiny*, 285–86.
50. Marshall, *Evolution*, 296.

Directed Evolution

phological level, where it is only possible to move from one adaptation to another through functional intermediates, a DNA sequence does not have to be functional in order to survive and be reproduced. In addition, DNA sequences are not under any selective surveillance while in "evolutionary transit" over succeeding generations. Denton writes: "Thus, new organs and structures that cannot be reached via a series of functional morphological intermediates can still be reached by a change in DNA sequence space." We could therefore say that DNA is eminently fit for directed evolution on account of the following: (i) the proximity of all life forms at DNA level, and that all known sequences can be inter-converted in small natural steps; (ii) the unidirectional flow of information from genotype to phenotype; and (iii) since functional DNA sequences can be derived via functionless intermediates, a new phenotype or organ can be generated by saltation, i.e., a mutational "jump."[51]

The vital importance of DNA to an organism is evident from the fact that it contains all the information necessary for the growth of the organism, which occurs by means of cellular multiplication and differentiation acting through the regulated expression of this information. Perry Marshall writes that DNA does not only *resemble* code, it *is* literally code. Since the codons on DNA strands are encoded into messenger RNA and decoded into amino acids and proteins, the genetic code thereby meets the requirements for a communication system, namely a code, an encoder, a digital message, and a decoder.[52] And since, according to Michael Denton, genes direct all of the morphological development of an organism from egg cell to adult, there is in principle no reason why genes cannot also direct evolutionary change. During the process of development the genes are rearranged at precisely predetermined times, as we see with the development of the immune system. Thus, "most genetic change underlying evolution, especially in higher organisms, has been largely a matter of the rearrangement of pre-existing genes rather than the emergence of new genes." In addition, variations in the sequence of the base pairs A-T and C-G can be

---

51. Denton, *Destiny*, 278–79.
52. Swift, *Evolution*, 306; Marshall, *Evolution*, 55–56.

explained by directional mutational pressure and selective constraints which are inherent in the genome itself.[53] Apparently the genetic data acts as efficient cause in both genotypical and phenotypical development.

However, we contend that to attribute the entire evolution of the living kingdoms to DNA programming alone is erroneous. The cellular and molecular biologist Martinez Hewlett has observed that a number of surprising discoveries were made during the course of the human genome project, which lasted from 1986 until 2001. First, there are large regions of DNA which, as they contain no genes, are therefore not involved in protein sequencing. Second, the total number of human genes amount to between 20,000 and 25,000, which is not much different from the total genes of the fruit fly. Third, it was found that genes are not definable in terms of their DNA sequence only. These results appear to confirm the argument by Michael Polanyi that, in principle, life cannot be reduced to the chemistry of DNA. Thus did molecular biology finally arrive at the conclusion that genetic information is irreducible to chemistry, thereby thwarting the project by Francis Crick and others to explain everything biological in terms of physics and chemistry.[54]

The following genetic evidence is presented by Michael Denton in support of molecular direction. First, the rate of evolutionary substitution, i.e., the average number of changes in DNA per generation since two species separated, and the rate of mutation, i.e., the changes in DNA sequence over one generation, are almost identical. This implies that different DNA sequences between species are generated by mutation, so that other factors such as natural selection play a minor role. Neo-Darwinists have tried to explain this phenomenon by means of the so-called "junk" hypothesis, which was first postulated in 1972 by geneticist Susumu Ohno and then popularized by Richard Dawkins in his work *The Selfish Gene* (1976). Since only 3 percent of the human genome codes for proteins, it means that 97 percent of DNA had no known function and

---

53. Denton, *Destiny*, 279–81.
54. Hewlett, *Biology*, 180–81.

Directed Evolution

was therefore viewed as non-functional evolutionary garbage. Remarkably, this "silent" DNA comprises the bulk of the DNA in all higher plants and animals.[55] But there are many challenges to the "junk" hypothesis. For one, it has since been discovered that in the human genome there are at least 4 million gene switches that reside in the non-functional DNA and control cellular behavior. Then, it is known that messenger RNA comes not only from protein-coding genes but also from many other parts of "junk" DNA. In addition, some of the so-called "junk" DNA contains a large quantity of transposon DNA, which are the so-called "jumping genes" that leap from one chromosome to another, thereby contributing functionally useful diversity to genomes. This implies non-random behavior, with transposons acting as coding sequences that mutate differently from other sequences. Moreover, there is evidence that non-protein coding "junk" DNA has informational characteristics similar to those of human language.[56]

Denton also argues that the uniformity of molecular evolution is established by comparing gene sequences. For example, all the higher cytochromes from yeasts and plants to insects and mammals display an equal degree of sequence divergence from the bacterial cytochrome in *Rhodospirillum*. This implies that all their cytochrome genes could have evolved at a uniform rate. The same uniformity is found in protein-coding genes, with very few exceptions. As Denton aptly remarks, it is difficult to explain this "molecular clock" in selectionist terms. For instance, human and salmon hemoglobins are equidistant from those of the hagfish, a living fossil having remained largely unchanged for around 300 million years, in spite of the enormous morphological and physiological differences between the human and salmon organisms. These genetic discoveries provide indirect evidence that the mutational processes which are changing the DNA sequence of organisms are being directed.[57]

---

55. Denton, *Destiny*, 288–89; Marshall, *Evolution*, 273; Swift, *Evolution*, 131.

56. Marshall, *Evolution*, 273, 294; Alexander, *Creation*, 59–60; Denton, *Destiny*, 290.

57. Denton, *Destiny*, 290–92.

Evolution by Means of Adaptive Mutation

In his previously-mentioned work, *Evolution 2.0: Breaking the Deadlock between Darwin and Design,* Perry Marshall presents compelling evidence from molecular biology in favor of goal-directed evolution, contrasting it with the Neo-Darwinian thesis of "random copying errors" driving the evolutionary process. Marshall explains, "Half the thesis of this book is that randomness does not create codes; and that once they exist, randomness can only destroy them. The other half of this thesis is that the origin of life required the creation of codes, and that nonrandom, linguistic adaptations of DNA continue to create codes and thus drive biological evolution." The following mechanisms are involved in this adaptive process as it occurs on the cellular level: transposition, horizontal gene transfer, epigenetics, symbiogenesis, and genome duplication (or hybridization). Remarkably, but not unexpectedly, these are all but ignored by Richard Dawkins in his voluminous celebration of the Neo-Darwinian theory titled *The Greatest Show on Earth* (2009).[58] We will now briefly consider Marshall's discussion of these evolutionary mechanisms.

Transposition is the rearrangement of cell DNA in response to external stimuli. That this does not take place randomly but according to precise rules was discovered by the cytogeneticist Barbara McClintock in the 1940s. McClintock experimentally subjected large numbers of maize plants to radiation and found that a minority of the radiated plants activated previously latent parts of the maize genome in order to "patch" damaged DNA with a new, transposable genetic element. By means of transposition, the maize plants changed the expression of their genome to meet an environmental challenge, in this case, radiation. McClintock's results were later confirmed by the geneticist James Shapiro, when he found that bacteria are also able to transpose elements of their DNA. In addition, it has been discovered that protozoans restructure their genome to create a new cell nucleus. This is done within a few hours in response to stimuli such as heat shock, pollution, hazardous chemi-

---

58. Marshall, *Evolution,* 149, 282.

Directed Evolution

cals, absence of food, and presence of indigestible food. In this way cells adapt purposively, and not randomly, to environmental pressures. This capacity is of vital importance to all of us, since our immune cells rearrange their DNA in order to produce specific antibodies to combat specific invaders. The implications for evolutionary theory are highlighted by recent studies showing that around 20 percent of the differences between placental mammals and marsupials could be attributable to large-scale rearrangements of genetic data by means of transposition.[59]

In horizontal gene transfer, cells exchange DNA with other cells. This activity negatively affects vast numbers of people every year in the form of bacteria developing resistance to antibiotics. When antibiotics are administered to a patient, the affected bacteria search for useful cells nearby. One such cell provides a purging "pump" that can make the bacteria immune to the antibiotic. By means of a plasmid transfer, the DNA that codes for such a pump is then transferred to non-resistant bacteria, enabling the latter to become antibiotic-resistant. They quickly multiply, so that this new and useful genetic material is distributed throughout the bacteria population. This process of adaptive mutation can take less than half an hour to accomplish.[60]

Horizontal gene transfer is facilitated by the ability of cells to communicate with each other and edit their own genomes by means of sophisticated language. Contrary to an earlier view popularized by Ernst Haeckel, cells are not formless masses of protein but are actually highly complex in both morphology and behavior. For instance, bacteria (which are the most numerous kind of cells on Earth) communicate with each other by sending out minute molecules which act in the same way as letters in human languages. Thus, Marshall writes, "different words form commands or requests, which are understood by their neighbors. Each molecule in the chain acts as a letter in the word." It further appears that the language of bacteria is not only chemically based, but also, like human language, based on the context of the situation. Astonishingly, bac-

---

59. Ibid., 82–83, 85, 89–90, 295.
60. Ibid., 94–96.

teria even appear to be multilingual, communicating in a native language for their own species and in foreign languages for other species.[61]

Epigenetics (Greek *epi*, above or upon + genetics) involves cells switching genetic code on and off for themselves and for their offspring and occurs in response to environmental challenges such as scarcity of food. Although the coding sequences in the cells remain the same, their expression is altered through a combination of mechanisms. As an example, cells regulate their energy consumption by storing more fat in times of abundant food, or use fat as a nutrient in times of food scarcity. A further example is a tree blooming on a particular day in the spring, but not again on an identical warm day in the autumn. This is due to its cells having recorded their memory of winter, and passing on this memory to the offspring of the tree. Plants are able to coordinate their cellular activities with a complex network of electrical, chemical, genetic, and pheromone signals. In this way they respond epigenetically to environmental signals by initiating subtle heritable changes to their offspring. Marshall notes that in contrast to transposition and horizontal gene transfer, "Epigenetics is primarily a tool of fine-tuning and gradual adaptation over multiple generations."[62] We would add that epigenetics is an important mechanism in micro-evolution, which occurs gradually and continuously over long periods of time.

A significant implication of epigenetics is that the evolutionary theory of Jean-Baptiste Lamarck has been vindicated, at least to some extent. As we noted in an earlier chapter, Lamarck was the French naturalist who published a theory of organic evolution 50 years before Darwin. Lamarck's view that organisms can pass on acquired characteristics to their offspring was partially accepted by Darwin, but was eventually discredited by the genetic work of Weismann and others. These founders of genetics strove to demonstrate experimentally that an insurmountable barrier exists between an organism's reproductive cells and its bodily cells, so that changes in

---

61. Ibid., 108–09, 113.
62. Ibid., 115, 117–19.

Directed Evolution

the latter do not affect the former and thus its offspring. Marshall notes, however, that in 2009 the *MIT Technology Review* reported the findings of experiments on rat pups as follows: "The effects of an animal's environment during adolescence can be passed down to future offspring.... The findings provide support for a 200-year-old theory of evolution that has been largely dismissed: Lamarckism, which states that acquired characteristics can be passed on to offspring."[63]

The process whereby cells merge and cooperate with each other is called symbiogenesis. The term is derived from the Greek words *syn* (together), *biōsis* (living) and *genesis* (origin)—in other words, the origin of living together. Symbiogenesis occurs not only among individual organisms as bees and flowering plants, but also on the cellular level in all plants and animals. The concept was first developed by the Swiss botanist Simon Schwendener in 1867, with his hypothesis (since verified) that algae and fungi had in the past merged to form a new species, namely lichen. This merger enabled lichen to survive in extreme environments where neither algae nor fungi could survive on their own. The first scientists to describe symbiogenesis in detail were the Russian botanists Konstantin Mereschowsky and Boris Kozo-Polyansky, during the early decades of the twentieth century. By the 1960s the theory had been revived by other scientists, notably the American biologist Lynn Margulis. In a more recent work, *Acquiring Genomes: A Theory of the Origins of Species* (2003), Margulis contends that evolution is primarily driven by symbiogenesis, which produces new organelles, bodies, organs, and species. The major classes of cells, plants, and animals are therefore constructed from symbiotic mergers of multiple smaller organisms, which explains their near-identical DNA. Margulis also notes the absence of experimental evidence that random mutations can bring about inherited variations. Contrary to the Darwinian insistence on competition as driving evolution, it now appears that cooperation is the primary driving force in producing beneficial variations. Natural selection therefore only functions

---

63. Quoted in Marshall, *Evolution*, 119.

after the appearance of genetic novelty, which is produced by the "creative forces" in DNA, such as symbiogenesis.[64]

Genome duplication occurs during the process of hybridization, when individuals from different species have mated in order to produce a new species. It was first introduced by the geneticist Susumu Ohno and some of his colleagues in *Evolution by Gene Duplication* (1970), in which they postulated that genome duplications or doublings took place during the early vertebrate lineage, thereby producing sudden and radical transformations of body plans. Studies of genome sequences suggest that around 500 million years ago, two species of sea squirt or tunicate mated to produce a new species, the hagfish. This first jawless vertebrate had 28 chromosomes, twice the number of its tunicate parents. It is unknown how long it took for the hagfish body plan to become established, Marshall admits, but apparently the genome itself doubled in one generation. Around 50 million years later, a second genome duplication took place when two jawless vertebrates mated to produce the first jawed vertebrate. This new species had 56 chromosomes, double the number again of its parents, and was an early ancestor of the vertebrate classes of bony fishes, amphibians, reptiles, birds, and mammals.[65] We see that these crucial morphological transitions occurred rapidly and not gradually, thereby lending further weight to the theory of punctuated equilibrium as postulated by Stephen Jay Gould and Niles Eldredge in 1972.

According to biologist James Shapiro, hybridization is more likely to occur after an environmental crisis like a forest fire or an earthquake, when organisms unable to find a mate within their own species would be spurred to mate with a partner from a different species. As a matter of fact, Marshall writes, genome duplication through hybridization has been observed between wheat and rice, butterflies and moths, and donkeys and mules. Moreover, further genetic research has suggested that among fish families such as salmon and carp, a third round of genome duplication has taken

---

64. https://en.wikipedia.org/wiki/Symbiogenesis; Marshall, *Evolution*, 122–23, 126, 130.

65. Marshall, *Evolution*, 135, 137–39.

Directed Evolution

place, thus providing them with eight times the number of chromosomes of the original invertebrates.[66]

All of these systematic mechanisms of adaptive mutation obey rules, Marshall emphasizes, and can be reproduced experimentally. However, a significant qualification has to be made: whereas transposition, horizontal gene transfer, and epigenetics are a matter of stepwise, "continuous improvement" adaptations, the remaining mechanisms of symbiogenesis and hybridization achieve "quantum leap" results.[67] We contend that this distinction relates precisely to micro- and macro-evolution respectively, micro-evolution being driven by cellular adaptations such as transposition, horizontal gene transfer, and epigenetics, while macro-evolution is driven by the mutational "jumps" provided by symbiogenesis and genome duplication.

In the light of the evidence above, Marshall concludes that cells are capable of doing their own genetic engineering, so that natural genetic engineering and not natural selection becomes the driving force of evolution. Consequently, evolution is neither slow nor gradual, but fast; not accidental, but organized; not purposeless, but adaptive. The Neo-Darwinian formula of random mutation + natural selection + time = evolution should therefore be replaced with the more accurate formula of adaptive mutation + natural selection + time = evolution. That is to say, the evolutionary process is directed by means of adaptive mutations on the cellular level, with natural selection acting afterwards as a sifting mechanism. Evolution is accordingly defined by Marshall as "the cell's capacity to adapt and to generate new features and new species by engineering its own genetics in real time."[68]

Organic Plenitude

In his work on directed evolution, Perry Marshall interprets the Divine command to Adam and Eve in Genesis 1:28 in terms of molecular biology: "Evolution is ultimately driven by cells' desire to

---

66. Ibid., 139–40.
67. Ibid., 247, 294.
68. Ibid., 89, 144–45, 173.

multiply, to fill the Earth, to use every available resource to its maximum potential, and to populate every ecological niche with fantastic beauty and diversity." Accordingly, any unbiased observer of the natural kingdoms has to be struck by the remarkable diversity of life on Earth in spite of the existence of embryological, physiological, and evolutionary constraints. As a matter of fact, Michael Denton notes, environmental constraints, such as a constant source of energy and chemical and physical stability, are important factors in the organic realm. The hydrosphere of the Earth satisfies these conditions through geo-chemical cycles, for example. The Earth also contains a large variety of environments, from polar oceans and high mountains to tropical forests and hot deserts. It is thus ideally suited for the manifestation of a plenitude of life-forms. The phenomenon of biodiversity supports the notion that the evolutionary tree of life on Earth was generated by direction from a unique program embedded in the order of nature.[69]

This plenitude of the organic realms had already been noted in 1835 by the French naturalist Georges Cuvier, who wrote that "Nature, inexhaustible in fecundity and omnipotence ... has realized all those combinations which are not incoherent."[70] And the related notion of an evolutionary tree of life has been vividly depicted by Charles Darwin in *Origin of Species*:

> The affinities of all the beings of the same class have sometimes been represented by a great tree.... The green and budding twigs may represent existing species; and those produced during each former year may represent the long succession of extinct species.... As buds give rise by growth to fresh buds, and these, if vigorous, branch out and overtop on all sides many a feebler branch, so by generation I believe it has been with the great Tree of Life, which fills with its dead and broken branches the crust of the earth, and covers the surface with its ever branching and beautiful ramifications.[71]

---

69. Ibid., 112; Denton, *Destiny*, 319–20.
70. Quoted in Denton, *Destiny*, 301.
71. Darwin, *Origin*, 100–01.

Directed Evolution

However, Marshall points out an interesting implication of horizontal gene transfer—that this well-known illustration of such an evolutionary tree should be revised. Thus, instead of a vertical tree with unconnected branches, demonstrating that all organisms have received their genes from their parents and more distant ancestors, it would depict a vast web of thin threads linking the branches to show the horizontal exchange of genes among living organisms.[72] Such an illustration would also reflect the phenomenon of evolutionary convergence (which we will discuss in the next chapter), as opposed to the divergence proclaimed by Darwinism.

According to the principle of plenitude obtained from Hellenic philosophy, as noted earlier, the actual existence of the universe must be as abundant as the possibility of existence, which is to say that no genuine potentiality of being can remain unfulfilled. This ontological notion is manifested *par excellence* among microbial life, since bacteria are found in virtually every ecological niche on Earth. Also, in their biochemical diversity, bacteria apparently utilize every available reaction in energy metabolism.[73] In fact, bacteria could be said (as was done by the biologist Carl Woese) to represent a third kingdom of living beings, next to plants and animals. Woese suggested that during the eons when life on Earth consisted mostly of single-celled organisms such as bacteria, evolution occurred mainly by means of horizontal gene transfer and not through a series of accidental mutations. The remarkable linguistic abilities of bacteria have been touched upon in the previous section. It is therefore not surprising to learn that bacteria are highly social organisms, similar in behavior to ants or bees. For example, they greet each other, search for food together, pool their digestive enzymes, and rearrange their genomes to survive environmental changes. A striking example of the latter is the *Anabaena*, a genus of cyanobacteria living as plankton. These bacteria divide labor by forming cells known as heterocysts, which are able to re-engineer their genomes to form an enzyme that extracts single nitrogen atoms from the air in cases

---

72. Marshall, *Evolution*, 98–99.
73. Denton, *Destiny*, 302–03.

of nitrogen deprivation. In other words, Marshall concludes, cells direct their own evolution.[74]

Denton mentions various examples of evolutionary constraints where all possibilities have apparently been realized: (i) the hibernation strategies in butterflies, namely egg, caterpillar, pupa or adult; (ii) the incubation strategies of birds; (iii) the structural possibilities of antelope horns and Gastropoda shells; and (iv) the modes of organic movement, i.e., jet propulsion in air and water, gliding, flying, ballooning, walking, swimming, and by propeller in some bacteria. And among the ubiquitous viruses, all conceivable ways of storing genetic information (RNA and DNA, each single- and double-stranded) are found, as well as the only possible forms of hollow structures that can be constructed from single subunits, namely the cylinder and the icosahedron.[75]

Regarding the vital bodily function of oxygen delivery, Denton notes that there are only two possible ways of delivering oxygen to the tissues of a terrestrial air-breathing organism: a circulatory system to convey dissolved oxygen via small tubes throughout the body, or a tracheal system of tubes to convey oxygen directly to the tissues. These alternatives are in fact employed by vertebrates and arthropods, respectively. The circulatory and tracheal systems of oxygen delivery are functionally (for example in terms of movement) best suited for larger and smaller organisms, respectively, and are thus not accidental. Moreover, these two systems are determined by physical constants such as the diffusion rate of oxygen in air and water, the density and viscosity of water, and so forth. As far as vertebrate lungs are concerned, there are again two different possible types: a bellows type in which air is inhaled and exhaled via the same passage, and a continuous throughput type in which air is inhaled and exhaled through different passages. These alternatives are actualized in terrestrial vertebrates and birds, respectively.[76]

In his *Critique of Judgment*, Immanuel Kant wrote that the analogy between a large variety of animal forms, for example in their

---

74. Marshall, *Evolution*, 96, 110–11.
75. Denton, *Destiny*, 305–06.
76. Ibid., 310–12.

Directed Evolution

skeleton, suggests conformity to a general original plan (German *Urbilde*) from which all are descended. It has been noted in this regard that the main body plans (German *Bauplanen*, literally "building plans") of the animal kingdom appeared hundreds of millions of years ago. These were subsequently to change, but new plans were not produced.[77] There appears to be a common "model" that provides a measure of structural continuity to the immense diversity of animal forms, although it is more apparent among animals possessing a higher level of consciousness, such as birds and mammals. This finds expression, for example, in the symmetrical disposition of the body, the number of extremities and sensory organs, and the general form of the internal organs.[78] However, David Swift notes an interesting non-homology pertaining to skeletons, namely that the vertebrae are embryologically formed in diverse ways for the different vertebrate classes. In birds, for example, the vertebrae are formed through resegmentation of the somite pairs, which does not occur in mammals; in some amphibians, the vertebrae develop from undifferentiated myotomal cells; in cartilaginous fish such as sharks and many of the "primitive" bony fish such as sturgeons, the vertebrae are formed from pairs of somite cells called arcualia; and in teleost fish, which comprise most living species, the vertebrae develop in three distinct stages totally unrelated to those of the other classes.[79] Clearly, the notion of directed evolution does not preclude morphological variety.

Within these main body plans there are two radically different types of skeletal structure, namely endoskeleton and exoskeleton. As in the breathing systems mentioned above, these skeletal types are actualized in vertebrates and arthropods. While an endoskeleton is better suited for larger animals, an exoskeleton is more useful to smaller animals. Also related to bodily size, the six legs of insects provide more stability during movement for smaller organisms than the four legs of vertebrates. Denton notes that there are a variety of ways to move on legs: humans walk and kangaroos hop on

---

77. Berg, *Nomogenesis*, 156; Popov, *Constraints*, 213.
78. Burckhardt, *Cosmology*, 144.
79. Swift, *Evolution*, 323–24.

two legs, most mammals walk or run on four legs, insects on six legs, spiders on eight legs, and centipedes and millipedes on more than ten legs. It thus appears that two basic designs for terrestrial macroscopic life exist: a larger type with a circulatory system, lungs, heart, endoskeleton and four legs; and a smaller type with a tracheal system, exoskeleton and six legs. These basic designs are actualized in vertebrates and insects, respectively.[80]

An interesting anomaly has been noted by Douglas Dewar, namely the ratio between types and species in various large groups of animals and plants. The British-Indian ornithologist reasons that if types and species are formed through the accumulation of variations or mutations as Darwinism asserts, then a group rich in types (i.e., morphological variety) should invariably contain more species than a group poor in types. However, the opposite is often the case, so that the number of species in the group is in inverse proportion to the number of types it contains. For instance, crustaceans and mammals are rich in types and poor in species, whereas insects and birds are poor in types and rich in species. The same rule is found in the plant kingdom: whereas the greenbrier (Smilaceae), rose (Rosaceae) and lily (Liliaceae) families are rich in types and poor in species, the aster (Compositae), true grass (Gramineae) and legume (Leguminosae) families are rich in species and poor in types.[81]

Despite the phenomenon of organic plenitude, gaps are also found in the *scala naturae* or ladder of being. For example, no intermediates are found between the unicellular Protozoa and the primitive multicellular Metazoan groups such as Porifera (sponges), Coelenterata (jellyfishes) and Ctenophora (comb jellies). There is an enormous diversity in morphology and behavior among both the Protozoa and these named groups, yet the gap between them is filled with only a few types of simple colonial Protozoa. This could be a necessary and not an accidental gap, Denton suggests, since simple life forms can be composed of one or many cells, but not of five or six cells. Other gaps in the ladder of nature could be due to different lifestyles: for example, the lack of marsupial whales or seals

---

80. Denton, *Destiny*, 312–14.
81. Dewar, *Illusion*, 243.

could be attributed to the marsupial reproductive system being unsuitable for an aquatic lifestyle. We may also observe that many of the seventy major phyla have never generated large complex forms and conclude that since the size of the flatworm is limited the absence of a circulatory system, every cell has to be near enough the body surface to receive oxygen.[82]

Remarkably, the immense morphological variety among major phyla such as the vertebrates, arthropods, and mollusks is not reflected in their DNA sequence space. Michael Denton postulates that this is because evolution is much easier to conceive of in DNA sequence space than in phenotypical space, just as it is easier to visualize movement on a two-dimensional map than in three-dimensional space. Accordingly,

> From the DNA perspective the whole evolutionary tree of life is in essence nothing more or less than a vast set of closely related DNA sequences clustered close together in the immensity of DNA sequence space, where each individual sequence is capable of specifying a viable life form, and where all sequences are interrelated and ultimately derivable via a series of steps from an original primeval sequence, which was the genome of the first life form on earth.[83]

We therefore contend that the unity of the organic world is genetically based, while its variety is due to environmental influences.

The Constraints of Functional, Integrative, and Irreducible Complexity

Evidently there are limitations on the principle of plenitude. Since all the parts and organs of an organism are functionally interrelated, constraints on one organ system necessarily impose constraints on others. As Denton explains, "The need for functional integration and coherence if an organism is to be actualized is bound to restrict the functionally possible to a fantastically small subset of all conceivable organisms." In other words, the functional integrity of organisms also imposes constraints on the evolutionary

---

82. Denton, *Destiny*, 314–15.
83. Ibid., 278.

pathways. Furthermore, evolution can only proceed through functional intermediates, which is another constraint on the possibilities of life-forms. This implies that perhaps not all possible fully functional life-forms can be generated. In addition to this evolutionary constraint are those associated with the process of development, namely embryological constraints such as the need for sufficient oxygen and nutrients, which require cardiovascular and respiratory systems. An example thereof is the gradual transformation in the embryo of a simple contractile tube into the four-chambered heart of the adult mammal.[84]

Due to the observed functional integrity of organic systems, living organisms are holistic in nature. It therefore follows that any variation to one of its parts has to effect most or all of the other parts, and therefore compensatory variations are required to preserve the organism as a properly functional entity. This presents a major obstacle to the Darwinian notion that all organic systems have arisen through the gradual accumulation of slight changes. As Darwin admits in *Origin of Species*, "To suppose that the eye, with all its inimitable contrivances for adjusting the focus to different distances, for admitting different amounts of light, and for the correction of spherical and chromatic aberration, could have been formed by natural selection, seems, I freely confess, absurd in the highest possible degree." However, in the same paragraph he then suggests that if numerous gradations from a complex eye to a simple eye could be shown to exist, and if such variations are both useful to the animal and inherited, then the creative power of natural selection would be vindicated.[85]

The argument of integrative complexity as presented by Stuart Kauffman, Michael Denton and other scientists states that "living systems are such intensely integrated systems that their components cannot be easily isolated and changed independently. Consequently, change, even if relatively minor, involves complex compensatory changes."[86] This Aristotelian notion of a whole presupposed by all

---

84. Ibid., 316–18.
85. Ibid., 328; Darwin, *Origin*, 143–44.
86. Denton, *Destiny*, 328.

its parts was expounded in the early nineteenth century by the influential French naturalist Georges Cuvier in his works translated as *Essay on the Theory of the Earth* and *Revolutions of the Surface of the Earth*. Cuvier argues that since every living being forms a single system consisting of mutually interacting parts, any changes to its parts require analogous modifications to the other parts. Since a living system is functionally integrated, any change in any of its subsystems would require highly specific and simultaneous compensatory changes to preserve its functional integrity. Denton thus concludes that "gradual changes resulting from the accumulation of a succession of minor independent changes is impossible."[87] And according to the biologist Stuart Kauffman, the reality of integrative complexity represents a fundamental constraint against adaptive evolution. In his *The Origins of Order* (1993), Kauffman writes that "as systems with many parts increase both the number of those parts and the richness of interactions among the parts, it is typical that the number of conflicting design constraints among the parts increase rapidly ... conflicting constraints are a very general limit in adaptive evolution."[88]

A related argument, which has become popular among Intelligent Design theorists, is that of irreducible complexity. In his work *Darwin's Black Box* (1996), the biochemist Michael Behe uses the bacterial flagellum to argue against the Darwinian theory of evolution. Bacteria possess a flagellum that enables them to swim, as we see, for example, with a mammalian sperm cell that propels itself through the female reproductive tract. In some cells the flagellum also serve as a sensory organelle, detecting chemicals and temperatures outside the cell. Remarkably, this minute apparatus consists of dozens of interdependent precision parts, so that the absence of any one of them would render the entire flagellum dysfunctional. The implication for evolutionary theory is that the bacterial flagellum could not have evolved by means of "numerous, successive, slight modifications" as Darwin claimed.[89]

---

87. Ibid., 329.
88. Quoted in Denton, *Destiny*, 329–30.
89. https://en.wikipedia.org/wiki/Flagellum; Marshall, *Evolution*, 171.

Behe's argument has been strongly criticized. However, as Perry Marshall points out, "none of the papers that challenge the irreducible complexity argument about the flagellum solve the problem within the traditional gradual-mutation framework." Moreover, critics of the irreducible complexity argument fail to recognize that cells are modular, which enables them to rearrange blocks of their genetic code by means of transposition and horizontal gene transfer, as mentioned earlier. These mechanisms, combined with the proven cognitive and linguistic abilities of cells, could have enabled the bacterial ancestors to assemble the flagellum as it exists today. Marshall concludes that the bacterial flagellum could indeed have evolved from simpler, pre-existing modular components, but not through a vast series of lucky steps as Darwinism asserts. Instead, through the various mechanisms of adaptive mutation, "it was built in a very few generations by a very smart cell. That bacterium wasn't lucky; it was *successful*."[90]

Even Richard Dawkins recognized (in *The Blind Watchmaker*) the existence of constraints that restrict or channel evolutionary change to some degree. But in true Darwinian fashion, he argues that in the case of a sufficiently large series of sufficiently finely graded intermediates, anything can be derived from anything else. However, Denton objects that this argument ignores both the functional constraints and the biophysical barriers to particular transformations, which no amount of intermediates can surmount. For example, for the reason mentioned earlier, all viral capsids are either cylinders or icosahedrons, and it is thus physically impossible for intermediates to function. The position of the nervous system in animals is also suggestive: ventral for invertebrates and dorsal for vertebrates. Due to integrative complexity, no group of organisms exist which have a nervous system halfway between the front and the back.[91] Here, again, we encounter a holistic reality, arising in the organic realm due to physical and biochemical laws regulating the unfolding of the One into the many.

---

90. Marshall, *Evolution*, 172–73.
91. Denton, *Destiny*, 330–32.

8

Convergent Evolution

IN THE HIGHLY INFORMATIVE and beautifully-illustrated *Encyclopedia of Animals* edited by Fred Cooke and other biologists, the term "convergent evolution" is defined as follows: "The situation in which totally unrelated groups develop similar structures to cope with similar evolutionary pressures." In his turn, the biochemist Denis Alexander defines convergence as "the repeated evolution in independent lineages of the same biochemical pathway, or organ or structure."[1] What is the relation of convergence to the well-known phenomena of parallelism and homology? Lev Berg asserts that there is no difference in principle between parallelism, convergence, and homology, since all organic characters arise in accordance with certain laws, or, more precisely, the actions of chemical substances affecting both morphology and physiology are subject to certain laws, and not due to infinite variations.[2] In other words, evolutionary convergence is a result of evolution according to natural law.

The phenomenon of convergence poses a problem for Darwinism, which emphasizes divergent evolution, so that homological similarities are ascribed to genetic relationship and convergence is viewed as rare and accidental. For example, Darwin argues in *Origin of Species* that the similarities of the bones in the hand of a man, the wing of a bat, the fin of a porpoise and the leg of a horse are due to descent with slowly acting successive modifications.[3] Berg, on the other hand, counters that in many groups of plants and animals the general trend of the evolutionary process is due to convergence,

---

1. Cooke et al., *Animals*, 588; Alexander, *Creation*, 326.
2. Berg, *Nomogenesis*, 159–60, 225.
3. Ibid., 360; Darwin, *Origin*, 361.

which affects both the external and internal characters of organisms. Since organic development is determined by laws, those organs are formed which must be produced according to the constitution of the organism and external effects, and are not random forms to be selected. Similarity of characters may therefore be due to a common origin, but it may also be the result of a certain uniformity in the laws of nature.[4] It has in fact been admitted by a leading Russian Neo-Darwinist, L. Tatarinov, that the reasons for the abundance of parallelisms remain unclear: "Random hereditary variation should cause a rather uniqueness of features, even in the case of the adaptation of related species to a similar environment."[5]

As we noted in an earlier chapter, the homology between structures such as the arm, fore-leg, wing and fin among various vertebrate classes was already understood by Aristotle. During the sixteenth century, naturalists such as Pierre Belon established the presence of homologous bones in the ostensibly very different skeletons of humans and birds. Eventually anatomists demonstrated the homology of not only skeletons but also organs among all the vertebrate classes. Even among arthropods as seemingly dissimilar as lobsters, flies, and butterflies, homologies are found in their exoskeletons.[6]

David Swift notes that homology, associated as it is with the Platonic notion of archetypes, is therefore a pre-Darwinian concept. However, Erasmus Darwin argued that the similarities among tetrapods (four-legged vertebrates) indicate their common origin, and it was in this sense that his grandson Charles drew homology into his theory as evidence of evolution. However, in Swift's view, the Darwinian view of homology as evidence of past evolution is problematic. For instance, the tetrapod limb which first appeared with the earliest amphibians is said to have evolved into such diverse structures as the bird's wing, the seal's flipper and the human arm, since they are all based on a common bone arrangement. But in each of

---

4. Berg, *Nomogenesis*, 360–61.
5. Quoted in Popov, *Constraints*, 215.
6. Dobzhansky, *Evolution*, 128; Berg, *Nomogenesis*, 158.

Convergent Evolution

the tetrapod classes—amphibians, reptiles, birds, and mammals—these bones have a distinctive arrangement and do not appear to be derived from a common underlying pattern. Moreover, in the fossil record there are no known gradations between these types of forelimb to show their evolution from a common ancestor. Instead, Swift observes that "the different animal groups, notably those with highly specialized limbs—such as pterosaurs, birds and bats for flying, and ichthyosaurs and cetaceans for swimming—appear abruptly, not progressively."[7]

Based on his extensive field-work on the fossils of soft-bodied animals around the world, Simon Conway Morris has become the leading thinker of our time on evolutionary convergence. Before embarking on a wide-ranging survey of convergence in his work *Life's Solution: Inevitable Humans in a Lonely Universe*, the British paleontologist notes that the building blocks of life are limited in number. These building blocks are: the nucleotide base pairs of DNA and RNA, containing genetic information and instructions; around 20 amino acids for building proteins; and the carbon-based sugars and fatty acids. From these relatively simple foundations arises all the vast diversity of the natural world. Conway Morris then adds that the notion of convergence provides a metaphor to describe how evolution navigates the combinatorial immensities of biological "hyperspace."[8] The beginnings of cosmic evolution—ostensibly humble, yet pregnant with potentialities—are affirmed by Denis Alexander: "The physical properties of the universe were defined in the very first few femtoseconds [i.e., one millionth of one billionth of seconds] after the Big Bang, and the process of evolution depends utterly on that particular set of properties." Thus in the organic realms, John Davison writes, convergence is "the selection of very similar morphologies drawn from a universal stockpile of preformed potentialities which were available when those evolutionary events took place."[9] Metaphysically speaking, it appears that over billions of years the unfolding of possibilities in the cosmos has

---

7. Swift, *Evolution*, 319, 321.
8. Conway Morris, *Life*, 27, 127.
9. Alexander, *Creation*, 135; Davison, *Manifesto*, 43.

proceeded from the physical and biochemical properties embedded in the ontological fabric of the whole.

Preparing the way, as it were, for convergence among plants and animals is the dynamic of molecular convergence, for which Conway Morris provides various examples. The protein molecule myoglobin, which is used to store oxygen, has apparently evolved independently in cyanobacteria and mammals. Moreover, high concentrations of myoglobin are found in the muscles of diving and burrowing animals, and an increased content of the amino acid arginine found in the myoglobin of both stands as evidence of molecular convergence. Two groups of fishes, the Arctic cod and the Antarctic notothenioids, generate anti-freeze proteins that are effectively identical, despite having different genes for producing these proteins. The enzyme lysozyme, which enables the digestion of tough plant material, provides a further example. The molecules of lysozyme have evolved convergently in such unrelated animals as cattle, certain Old World monkeys (e.g., the langur), and the enigmatic hoatzin bird of the Amazon forests.[10]

It is rather ironic that the advent of global warming, which is at least partially due to the human addition of vast quantities of carbon dioxide to the atmosphere, could well prove to be beneficial to the plant kingdom. Over roughly the past ten million years, until the advent of the industrial era, the levels of atmospheric carbon dioxide underwent a gradual decline, presenting a challenge to plants which require this gas for photosynthesis. In order to meet this environmental challenge, along with the challenge of increasing aridity and salinity, a number of plant species developed a new method, known as C_4 photosynthesis. Although biochemically highly complex, C_4 photosynthesis has evolved independently at least 31 times in the plant kingdom. This is notably true of grasses, which is highly significant in at least two respects. First, the development of extensive grasslands on most of the continents contributed much to mammalian evolution, especially of the herbivores (and thus also of the carnivores feeding on the herbivores). Second, grasses such as wheat,

---

10. Conway Morris, *Life*, 288, 290, 298.

rice and maize have been staple foods for the bulk of humankind over thousands of years, and have thus played a major role in human survival.[11]

Convergence of Internal Characters

In the case of the insect class, with its more than 30 orders, convergence affects all of the organs, according to Berg. For instance, in lowly organized insects such as Neuroptera (lacewings and ant-lions) there is a convergence of various characters with higher groups such as Diptera (flies) and Lepidoptera (butterflies and moths).[12] Berg adds that the similarity of vertebrates to invertebrate groups such as Annelida (segmented worms) is also due to convergence, whether they derive from one common stock or have polyphyletic origins. For example, there are structural resemblances in their urino-genital system, circulatory system, and olfactory organs. Moreover, the parallelism between Dipnoi (e.g., lungfishes) and Amphibia pertaining to their lungs, skull, brain, and pelvis shows that organisms develop in conformity to inherent forces, which, in the presence of certain conditions, become embodied in definite forms.[13] This reference to "inherent forces" determining organic morphology evokes the Hellenic doctrine on the priority of formal causality, without denying the reality of efficient and material causation.

When comparing dinosaurs and birds we again find a number of structural similarities, for example the existence of pneumatic bones even when the saurians did not fly; walking on the hind limbs, i.e., a bipedal motion; and an increase in the number of sacral vertebrae from three to ten, compared to the two or three in living reptiles. Also, the first digit in the foot of many dinosaurs is contraposed and reversed, as in birds, while in some saurians it is rudimentary or completely disappears, as is also the case in birds like the ostrich, plover, and auk. Berg concludes that since the flying saurians were not the ancestors of birds, their similarities in structure are

---

11. Ibid., 292–94.
12. Berg, *Nomogenesis*, 164–66.
13. Ibid., 167–68, 172–75.

due to convergence—that is to say, development in a determined direction.[14]

Berg discusses numerous other structural similarities between animals of unrelated groups, among them the remarkable range of resemblances between crocodiles and birds. These include a four-chambered heart, as well as the lungs, ribs, structure of the forelimbs, nervous system, hearing organs, and digestive organs. Even the brain of crocodiles resembles birds more than those of any other reptiles. Mammals and the flying saurians known as pterosaurs show marked similarities in their skeleton, for example the femur (thigh bone) and the humerus (upper arm bone). Berg points out that the habits of mammals and pterosaurs are so different that their similarities cannot possess any adaptive significance. More examples of convergence between unrelated groups include the following: the limbs and teeth of Equidae (the horse family) in North America and Litopterna (an order of extinct ungulates) in South America; Ichthyosauria (an extinct order of large marine reptiles) and dolphins, pertaining to the head, neck, fins, and so forth; and diurnal birds of prey and (nocturnal) owls displaying many points of resemblance in internal structure, for example the muscles of limbs and in the crop, although belonging to widely separated orders.[15]

Further examples of convergence in organs among unrelated animals are mentioned by Douglas Dewar in his anti-Darwinian classic. Among these are the tracheae that constitute the breathing organs of velvet worms, millipedes, centipedes, insects, spiders, scorpions, and wood-lice; and similar types of nephridia, which are the invertebrate equivalent of vertebrate kidneys, in amphioxus and polychaete worms. Structural convergence among unrelated taxa occurs likewise among plants, as in the case of monocotyledons and dicotyledons, possessing one and two seed-leaves respectively. These parallel, polyphyletic stems developed independently, Berg argues, so that their mutual similarities are due to convergence.[16]

The immense range of convergence in the organs of animals and

---

14. Ibid., 175–79, 185, 189.
15. Ibid., 187–88, 192, 212–13.
16. Dewar, *Illusion*, 39; Berg, *Nomogenesis*, 205.

plants sometimes illustrates the principle of phylogenetic acceleration. Thus the placenta, the organ connecting the embryo to its mother, was formed independently in Polyzoa (moss animals), *Peripatus* (velvet worm), certain insects and scorpions, certain marsupials, and all placental mammals. It is highly unlikely that this widespread phenomenon is due to common descent, since the mammalian groups of monotremes, marsupials, and placental mammals are parallel branches that arose independently of one another.[17] The emergence of a placenta is due to the retention of the embryo within the female reproductive tract, which is the technique of ovoviviparity or the giving birth to live offspring that have developed from eggs laid by the mother. This method of reproduction is estimated to have evolved independently around 100 times among lizards and snakes, as well numerous times among frogs and fishes. Moreover, the Brazilian lizard *Mabuya heathi*, a member of the skink family, converges on mammals in its reproduction, having a placenta, a minute egg comparable to a mammalian ovum, and a prolonged period of gestation. Some viviparous lizards, such as the European *Chalcides chalcides* or three-toed skink, even develop a placentome which is cast off when giving birth, similar to the mammalian afterbirth.[18]

The 70 species of the fascinating yet little-known velvet worm constitute the invertebrate phylum Onychopora, living in tropical and southern temperate zones of the world, including South Africa and New Zealand. With its combination of arthropod and annelid characters, the velvet worm represents one of the most remarkable instances of convergence in the animal kingdom. Incidentally, the velvet worm genus *Peripatus* also nourishes its developing embryos with a kind of placenta, as noted by Lev Berg. *Peripatus* thereby combines features of three major taxa: the phyla Annelida and Arthropoda, and the placental Mammalia.[19]

In view of the foregoing facts it is not surprising that sensory convergence is also widespread. The eye has evolved at least 20

---

17. Berg, *Nomogenesis*, 214–16.
18. Conway Morris, *Life*, 220–21.
19. Cooke et al., *Animals*, 587; Davison, *Manifesto*, 42.

times independently, says Conway Morris, and includes at least 15 lineages of photo-receptors with a distinct lens. That the availability of time cannot be an objection to such repeated evolutionary outcomes is evident from the finding by D.-E. Nilsson and S. Pelger that a simple eye-spot can be transformed into a fully functioning camera-eye in substantially less than a million years (which, we would add, suggests a series of macro-mutational "jumps"). It is well-known that the eye of vertebrates and of cuttlefish (related to squid) are structurally very similar, each possessing a lens and an iris with comparable performance. However, David Swift argues that these eyes are not actually homologous, since they arose independently in the different phyla of chordates and mollusks which have completely different body plans and phylotypic stages.[20]

Another optical convergence between vertebrates and invertebrates is the remarkable phenomenon of organisms with pairs of eyes for simultaneous vision in air and water. It occurs, for instance, in the Central American four-eyed fishes, namely the fresh-water *Anableps* and the marine *Dialommus*, which belong to different orders, and in whirligig beetles (*Gyrinus*), where it serves precisely the same purpose.[21] Interestingly, whirligig beetles count among the roughly 30,000 species of aquatic insects, all said to have evolved from terrestrial ancestors. However, Dewar objects that such a descent would have required drastic changes in anatomy and physiology pertaining to the modes of locomotion, feeding, and respiration, which have not been satisfactorily explained.[22] Optical convergence also occurs between chameleon lizards and sand-lance fishes, in which the camera-eye has been modified with the replacement of the lens by the cornea. The latter is fitted with muscles to enable focusing, which enables the chameleon and the sand-lance to rapidly strike at their prey, with the tongue and the whole body, respectively.[23]

It is noteworthy that the sense of sight is not restricted to ani-

20. Conway Morris, *Life*, 164, 387; Swift, *Evolution*, 320.
21. Berg, *Nomogenesis*, 219, 306–07.
22. Cooke et al., *Animals*, 562; Dewar, *Illusion*, 251–52.
23. Conway Morris, *Life*, 164.

mals, as Conway Morris points out. Photosensory devices such as eye-spots are found in a number of single-celled organisms, for example the eukaryotes *Chlamydomonas* and *Volvox*. Moreover, the visual protein rhodopsin employed by these eye-spots appear to be convergent on the equivalent protein found in human and other animal eyes. Also relevant is the complex optical system of the single-cellular organisms called dinoflagellates. In spite of being less than one tenth of a millimeter in length, their optical apparatus is strikingly convergent on the animal eye. In view of all this evidence, the evolution of eyes is viewed as virtually inevitable: "[T]hat key molecules required for vision, such as rhodopsin and the crystallin proteins, evolved in single-celled organisms ... suggests that, given time and the adaptive value of light discrimination, the evolution of the eye seems to be a near inevitability."[24]

Various instances of "seeing" without eyes are found among some mammal and bird species. For example, the star-nosed mole (*Condylura cristata*) in North America uses the extremely sensitive, fleshy tentacles on its nose to "see" while burrowing in soil or swimming in water, even though it is almost blind. Some animals have developed echolocation to "see" in the dark, the most famous being bats. Almost incredibly, certain moth species favored by bats produce ultrasonic clicks, apparently to jam the bat's sonar. Echolocation is also used by dolphins, as well as by birds inhabiting deep, dark caves, such as South American oilbirds and Asian swiftlets. As Conway Morris aptly remarks, with their faculty of echolocation, animals such as bats, dolphins, and some birds inhabit a sensory world that is completely unknown to humans.[25]

Regarding the sense of smell, or olfaction, we find evidence of convergence between vertebrates and insects, in spite of differences in their relevant anatomy (i.e., nose and antennae) and molecular genetics (i.e., different proteins to bind odors). For instance, the glomeruli of insects have precise counterparts in the olfactory bulbs of vertebrates, their olfactory sense operating on the same principles. And since the ancestors of vertebrates and insects, namely the

---

24. Ibid., 165, 173.
25. Ibid., 175, 177, 181–82.

amphioxus and aquatic crustaceans respectively, do not possess glomeruli, Conway Morris concludes that their practically identical olfaction arose independently.[26] We should keep in mind, however, that the postulated descent of vertebrates from invertebrate chordates such as amphioxus is disputed, as noted earlier.

As far as hearing is concerned, it appears that the ear with a tympanic membrane has evolved independently several times among the tetrapod or four-legged vertebrates. And as is the case with the other organs of sense, we find hearing convergence between vertebrates and insects. For example, in the mosquito a complex antenna serves as the equivalent of an ear. It includes an organ containing numerous nerve cells sensitive to sounds, so that the "hearing" of mosquitoes is remarkably similar to that of mammals. A tympanal structure analogous to the human eardrum is furthermore found among insects such as moths, grasshoppers, and crickets. Even the process of transduction, i.e., the conversion of a mechanical force such as sound into an electric signal, is similar in vertebrates and insects.[27]

Another fascinating bio-phenomenon is the electrical generation found in various species of fish. Freshwater species include the electric eel (*Electrophorus*) from South America and the electric catfish (*Malapterurus*) and elephant-nose fish (*Campylomormyrus*) from Africa. The former two species are capable of generating an electric current of several hundred volts for stunning prey and deterring predators. In its turn the elephant-nose discharges a weak electric current around its body, which is used for navigation at night or in murky waters. This faculty of electro-generation has evolved independently at least six times among fish of various genera, through modification of muscle cells. Electro-generation is also used for communication, each species producing a highly specific electric signal which enables the fish to "see" or "hear" in opaque water. Interestingly, in the African *Mormyrus* species the fish have an enormous brain in relation to their body size, comparable to the human

---

26. Ibid., 179–80.
27. Ibid., 190–93.

brain. This presumably facilitates sophisticated neurological processing in the dark, electrically charged water in which these fish live.[28]

There is also a physiological parallelism between animals and plants, which is not unexpected given the similar physical and biochemical constraints in their evolution. For example, in certain insectivorous plants a ferment similar to pepsin is secreted by corresponding organs for the digestion of animal food. Also, the chitin of animals such as annelids, mollusks, and arthropods is very similar in its chemical structure to plant cellulose. Another case of convergence due to chemical action is found between chlorophyll and hemoglobin molecules, both containing a nucleus approximating their pigments to the dye-stuffs of the indigo group.[29] The most well-known fauna of the Cambrian period are the trilobites, which roamed the world's oceans for over 270 million years until they became extinct at the end of the Permian. These highly successful arthropods had an exoskeleton which was biochemically similar to the cellulose found in plants.[30]

Similarity in organs found between different groups of organisms are sometimes due to common descent or genetics, and sometimes due to convergence. Lev Berg writes, "Both convergence and homology (similarity due to relationship) are governed by laws. Therefore, if evolution is nomogenesis, chance and natural selection evidently play no part in the origin of new organic forms."[31] Embryological evidence appears to support Berg's distinction between homology and phylogeny, against the Darwinian insistence that the similarities in homologous organs among unrelated species are due to a common ancestor, where, as David Swift remarks, homologous structures arise due to comparable embryological tissues as well as similar developmental processes. However, it has been discovered that certain apparently homologous adult structures do not arise from comparable embryological sources, as the eminent embryolo-

---

28. Cooke et al., *Animals*, 473, 485; Conway Morris, *Life*, 183, 185, 187, 194–95.
29. Berg, *Nomogenesis*, 223–25.
30. Swift, *Evolution*: 261; https://en.wikipedia.org/wiki/Trilobite.
31. Berg, *Nomogenesis*, 226.

gist Gavin de Beer admits in his *Homology: An Unsolved Problem* (1971). If similar morphological structures developed from different embryonic sources, as the evidence indicates, then homology can no longer be adduced as support for Darwinian evolution. As Swift writes, "the inconsistency between morphological homology and supposed phylogenetic relationships has become a further challenge to the theory of evolution."[32]

Convergence of External Characters

Berg attributes the phenomena of convergence and parallelism in external characters to inner or autonomic causes, not to the influence of the geographical landscape. After studying parallel variations in diverse cultivated plants, the botanist Nikolai Vavilov formulated the Law of Homologous Series of Variations: "Species and genera more or less nearly related to each other are characterized by similar series of variation with such a regularity that, knowing a succession of varieties in one genus and species, one can forecast the existence of similar forms and even of similar genotypical differences in other genera and species."[33] Such parallel variations are found in both cultivated and wild forms within a species, a genus, a family, or even further afield. Examples of this are found in (i) different species of wheat (genus *Triticum*), barley (genus *Hordeum*) and oats (genus *Avena*); (ii) different genera of the same family, such as watermelons, melons, cucumbers, and pumpkins displaying round, oblong, and flat shapes; and (iii) different families, such as the type of roots in the carrot, beet, and turnip families. Series of parallel variations are even observed in distant orders and classes, including the phenomena of albinism (i.e., lack of coloration), gigantism (i.e., excessive growth), nanism (i.e., dwarfism), and fasciation (i.e., crested growth).[34]

Darwin recognized many cases of "analogous or parallel variations," both in *Origin of Species* and *Variation of Animals and Plants under Domestication* (1868). However, he ascribed their origin either

---

32. Swift, *Evolution*, 330–31, 333.
33. Quoted in Berg, *Nomogenesis*, 236.
34. Berg, *Nomogenesis*, 237–40.

Convergent Evolution

to unknown causes having acted on organic beings with a similar constitution, or to the reappearance of characters possessed by a remote progenitor. Berg comments that the first explanation signifies nothing more than the unfolding of certain latent factors, which is to say that natural selection played no part. Against Darwin's notion of infinite variability, Berg argues that variations in the same direction will always ensue under definite conditions. In other words, the inherited tendency of organisms to vary in a like manner, thereby causing convergence, is not exceptional, as Darwin thought, but is actually a fundamental law of evolution in the organic world.[35]

That evolution is determined by law is the reason we observe many parallel forms in all of the vertebrate classes, Berg adds. Among mammals it has been observed that in both the primate tarsiers (genus *Tarsius*) and the marsupial colocolo (*Dromiciops*) the second and third toes have almost identical claws, which are used as a toilet comb. Interestingly, recent DNA studies have confirmed that the colocolo found in Argentina and Chile is more closely related to Australian than to other South American marsupials, thereby reinforcing the theory that marsupials migrated from South America to Australia via Antarctica between 100 and 65 million years ago, when these continents were joined together as the Gondwana landmass.[36]

A remarkable case of evolutionary convergence in action can be seen in New Zealand, which has been isolated in the Pacific Ocean for at least 85 million years. Prior to the arrival by boat of Polynesians around the year 1000, these islands were too remote to be colonized by terrestrial mammals, except for bats. Conway Morris notes that several species of birds and even insects have evolved converging on mammals. For example, the giant wingless crickets known as wetas (*Deinacrida heteracantha*) are related to locusts, but converge on mice and rats in biomass, nocturnal foraging, use of diurnal shelters, polygamy, and even their droppings. An endemic bat species has become partially terrestrial and walks on the ground with its wings folded. Most remarkable of all is the kiwi (genus

---

35. Ibid., 240–44.
36. Ibid., 244–45; Dewar, *Illusion*, 39; Cooke et al., *Animals*, 70.

Apteryx). This flightless, nocturnal bird has feathers that are fur-like, which facilitates living in burrows, its nostrils are located at the tip of the beak rather than the base, thus improving its sense of smell when foraging at night, and the feathers around its mouth are whisker-like. Even more convergent on mammals is the kiwi's protracted incubation period, although it is egg-laying. For reasons of both anatomy and behavior, the kiwi has, not surprisingly, been referred to as an honorary mammal.[37]

The most well-known instance of convergent evolution among entire organisms is the equivalence between a large number of placental mammals and their marsupial counterparts, both filling similar ecological niches. Marsupials or pouched mammals comprise seven orders, of which three are found in the Americas and four in Australasia. The fossil record indicates that marsupials and placental mammals diverged more than 100 million years ago. Examples of placental-marsupial convergence include the grey wolf (*Canis*) and the extinct Tasmanian tiger (*Thylacinus*); the ocelot (*Leopardus*) and spotted-tail quoll (*Dasyurus*); the giant anteater (*Myrmecophaga*) and numbat (*Myrmecobius*); the flying squirrel (*Glaucomys*) and glider (*Petaurus*); the groundhog (*Marmota*) and wombat (*Vombatus*); the house mouse (*Mus*) and mulgara (*Dasycercus*); and the golden mole (*Amblysomus*) and marsupial mole (*Notoryctes*), both being sightless and insectivorous. In addition, the koala (*Phascolarctos*) and the flying lemur (*Galeopithecus*) are the marsupial equivalents of the bear (*Ursus*) and the flying squirrel (*Pteromys*), respectively.[38]

Another interesting case of marsupial-placental convergence is the striped possum (*Dactylopsila trivirgata*) of Australia and New Guinea and the aye-aye (*Daubentonia madagascariensis*) of Madagascar, both of which are arboreal insectivores. Both of these unrelated species have an elongated finger that enables them to extract wood-burrowing grubs.[39] There is also a striking resemblance

---

37. Conway Morris, *Life*, 218–20.
38. Cooke et al., *Animals*, 68; Bergman, *Orthogenesis*, 143; Berg, *Nomogenesis*, 303.
39. Cooke et al., *Animals*, 68.

Convergent Evolution

between the extinct placental and marsupial saber-toothed cats, with their dagger-like teeth which developed independently. Their similarities extend even to the large flange on the lower jaw, which according to Otto Schindewolf was designed to guide and protect the upper canines.[40] To this impressive list of marsupial-placental convergence we would add the Tasmanian devil (*Sarcophilus harrisii*) as the marsupial equivalent of the African honey badger (*Mellivora capensis*, known in southern Africa as the ratel), in both morphology and behavior.

Among mammals, only bears and humans have a bipedal gait (apes do not qualify, since they have to use their arms for support). It is therefore not surprising to learn that in both bears and humans the first toe is elongated. Among rodents and insectivores, convergence is found in mice and shrews, and porcupines and hedgehogs, respectively. The aquatic mammalian orders Cetacea (whales and dolphins) and Sirenia (manatees and dugongs) are anatomically far apart but nonetheless possess various structural similarities. They all display a fusiform body, an absence of posterior extremities, a transformation of fore-limbs into flippers, the presence of a caudal fin, an elongation of the lung and its transformation into a kind of hydrostatic organ, and the disappearance of body hair and external ears. The transformation of hind-limbs into organs for jumping occurs in rodents, insectivores, and marsupials. In the South and Central American forests mammals of different orders are found with prehensile tails. These include a number of monkeys, the kinkajou (allied to bears), tree-porcupines, anteaters, and various marsupials.[41]

In the order Insectivora there are numerous cases of convergence among species that are not closely related, living in similar habitats on different continents. For example, the European desman and the Madagascar tenrec are both aquatic, with a waterproof coat, streamlined body, partially webbed feet, a long tail acting as a rudder, and specialized mechanisms for breathing and detecting prey underwater. European moles and African golden moles are burrow-

---

40. Alexander, *Creation*, 327; Davison, *Hypothesis*, 3.
41. Berg, *Nomogenesis*, 301, 303–05.

ers with cylindrical bodies, powerful short limbs, large digging claws, and tiny eyes hidden by fur or skin. European hedgehogs and African tenrecs both have a thick coat of spines, curling up into a spiky ball when they are threatened. Moreover, both Cuban solenodons and African tenrecs have apparently developed echolocation for finding their prey. There are also cases of convergent evolution due to a similar lifestyle among rodents; for example, tree squirrels and scaly-tailed squirrels are only distantly related, yet members of both families possess a membrane for gliding. Another example is the North American prairie dog and the South American degu, which both live in large colonies in extensive systems of burrows and communicate through a range of vocalizations.[42]

Among burrowing mammals, also known as fossorial species, evolutionary convergence occurs among three orders of 11 separate families and at least 150 genera, spread over all the continents except Antarctica. They converge in anatomy, physiology, behavior, and even aspects of genetics. In addition, some of the mole-rats such as *Heterocephalus* and *Bathyergus* have evolved a social system with one reproductive female, while the remaining rats are divided into castes, including workers. This colonial structure in which the breeding female, or queen, is protected from danger at all costs is found not only among mammals, but also converges in ants, bees, wasps, termites and certain beetles.[43]

In the case of birds, we observe convergence between the penguins of the southern hemisphere and the auks of the northern hemisphere. Both forms live in a cool oceanic environment and have wings modified into flippers for underwater swimming. However, they do not share a recent common ancestor, having been separated by the warm tropical waters for many millennia, and their similarities are therefore due to convergent evolution. Convergence between reptiles and amphibians is displayed by the subterranean lizards of the suborder Amphisbaenia and the amphibian order Gymnophiona, known as Caecilians. Both groups are built for burrowing, possessing a rudimentary tail, subcutaneous eyes, a soft skin

---

42. Cooke et al., *Animals*, 84, 218, 238.
43. Conway Morris, *Life*, 139–43.

deprived of scales, and a compact skull. In addition, both have a peristaltic motion of the body, as is also found in earthworms.[44] Regarding optical convergence, telescopic eyes with a supplementary retina are found in some deep-sea fishes, various crustaceans, and insects such as dragon-flies. The contrivance of claws for capturing prey is found in different invertebrate classes such as scorpions (Arachnida) and crayfish (Crustacea), as well as water insects such as water scorpions (family Nepidae) and diving beetles (family Dytiscidae).[45]

Among plants, Berg continues, we find convergence in the similar morphology of American cactuses, the African family Euphorbiaceae (spurges) and the South African succulent genus *Stapelia*, as a result of adaptation to definite external conditions. These plants, known collectively as xerophytes, live in conditions of extreme aridity and display a whole range of similarities. Conway Morris illustrates this using two unrelated desert plants, a Mexican cactus (*Peniocereus striatus*) and a Kenyan spurge (*Euphorbia cryptospinosa*). A cross-section of their stems shows a remarkable similarity in structure, both having flat ribs and intervening recesses for the preservation of water. Their similarity extends to the interior of the stems, with water-storing cells and a starch-rich central pith that provides food for the plant. Both species also possess a reddish pigment which provides a moribund appearance to the plant, presumably making it unattractive to herbivores.[46]

We also find structural convergence due to similar living conditions in Mediterranean-style floras in parts of the world as far from the Mediterranean as Chile, the Western Cape in South Africa, and Western Australia. It extends to plant-animal interactions such as the dispersal of seed by ants, of which convergence has been shown between Australian and South African species. Conway Morris writes that among plants from the tundra to the tropics, convergence is pervasive, for example in leaf structure. In the case of aquatic plants, Berg mentions the example of the dicotyledons of

---

44. Cooke et al., *Animals*, 26, 370, 427; Berg, *Nomogenesis*, 306.
45. Berg, *Nomogenesis*, 307, 310; Cooke et al., *Animals*, 562.
46. Berg, *Nomogenesis*, 311; Conway Morris, *Life*, 134–35.

the river-weed family Podostemaceae living in rapidly-moving water that are similar in both external form and internal structure to algae, liverworts, mosses, and lichens. As adaptations to aquatic life, the leaves of plants of various families are deeply dissected and become thread-like in shape, while some species have both submerged dissected and floating whole leaves.[47]

From the above-mentioned examples it is indisputable that convergence resulting from adaptation to a similar mode of life is ubiquitous. Berg concludes that this phenomenon is subject to certain laws and is therefore unaffected by natural selection.[48] In the meantime, the science of genetics has thrown further light on this aspect of convergence, as David Swift writes in his thought-provoking work *Evolution under the microscope*. For example, very different animal groups such as bony fishes, sharks, ichthyosaurs (extinct marine reptiles) and dolphins all have or had a streamlined body well-suited to their aquatic mode of life. The Neo-Darwinian explanation for this phenomenon is that numerous variations arose through random mutations in these groups, which were selected until the most suitable body shape emerged. However, the biochemical evidence is against the production of new genetic material during the evolutionary process, indicating instead that species have descended by segregation and selection from a primordial gene pool or that the unrelated aquatic fauna mentioned above acquired their similar streamlined forms not through random mutation, but because their primordial ancestors already possessed appropriate genes.[49]

Convergence of Psychical Phenomena

According to Lev Berg, instinct has not been derived from intelligence, nor vice versa, but both have presumably been independently derived from reflexes and have then developed along parallel lines. In accepting this, Berg appears to be following the dictates of scientific materialism, since intelligence cannot be reduced to an

---

47. Conway Morris, *Life*, 135, 365; Berg, *Nomogenesis*, 312–13.
48. Berg, *Nomogenesis*, 313.
49. Swift, *Evolution*, 250, 252.

epiphenomenon of the neuro-cerebral system. The Russian biologist stands on firmer ground when he asserts that the development of languages follow phonetic laws. For instance, each group of Indo-European languages developed new forms parallel to each other, which accounts for their common structural features. As an example, in many Indo-European languages a complex past tense is formed by combining a participle with an auxiliary verb.[50] Regarding linguistic convergence, experiments with dolphins have shown that they are able to master both syntax and semantics. According to the biopsychologist Lori Marino, species such as dolphins, chimpanzees, and perhaps African Grey parrots have converged to a level of cognitive complexity that is ready for articulate speech. Conway Morris therefore concludes that language is an evolutionary inevitability, or, according to the eminent linguist Noam Chomsky, it appears that the brain has a kind of grammatical software programmed into it.[51]

Among non-human mammals, the anthropoid apes and dolphins are generally viewed as the most intelligent species. In fact, some dolphin species have a substantially larger brain than the chimpanzee, gorilla and orangutan. According to the research of Lori Marino, dolphins were the biggest-brained organisms on the Earth until approximately 1.5 million years ago, when they were overtaken by the hominid *Homo erectus*. The large brain size of dolphins is related to their sophisticated social organizations and its concomitant advanced vocalizations, Conway Morris suggests. Recent studies have shown that dolphins have a more advanced social structure than the chimpanzees, which not only is genetically our closest animal relative but possesses complex and varied vocalizations, some of which converge on those of porpoises. According to Conway Morris, dolphin intelligence appears to be more convergent on humans than that of apes, as is evident from their complex social life, their sophisticated communication, their ability to learn and mimic, and also their playfulness and self-recognition. Despite the cetacean and primate lineages having diverged around 65–70

50. Berg, *Nomogenesis*, 260.
51. Conway Morris, *Life*, 252–53; Thompson, *Origins*, 43.

million years ago, and although strikingly different in neuro-anatomy, dolphins and the higher primates thus converge in cognitive ability.[52]

Another type of psychical convergence is the resemblance in social organization between two of the largest mammals, namely sperm whales and elephants. In both species, the females and the young form social units, with long-distance vocalizations and intense socialization. The males are solitary and wide-ranging, returning to the social units only when they are sexually mature and strong enough to compete in mating contests. Convergence between sperm whales and elephants can be seen in social complexity, communal care of the young, intelligence, memory and even longevity.[53] Convergence could even pertain to music—for example, the songs of male humpback whales are constructed according to the musical laws followed by human composers. This led the biologist Patricia Gray to relate whale and bird song to the Platonic theory of music: "The similarities among human music, bird song, and whale song tempt one to speculate that the Platonic alternative may exist—that there is a universal music awaiting discovery."[54] Here again we encounter the notion that the cosmos is based on law, as maintained by Plato and other Hellenic philosophers.

The concurrence of psychical convergence with social organization is displayed on a grand scale among certain invertebrate groups such as termites, ants, bees, and wasps. Such convergence occurs in considerable detail among different orders, for instance South American ants and African termites, with their fungus nurseries and differentiation between workers and soldiers. For example, the leaf-cutting ants in South and Central America (genera *Acromyrmex* and *Atta*) have invented agriculture independently of humans. The farming activities of these ants are centered around the harvesting of leaves for the fungus gardens in the ant nest. Further farming tasks for the ants include weeding, pruning, cropping, and even the use of herbicides to protect their fungus gardens against pests. These

52. Conway Morris, *Life*, 247, 249–50, 257–58.
53. Ibid., 250.
54. Quoted in Conway Morris, *Life*, 421.

industrious leaf-cutters form colonies of several million ants, displaying highly organized societies. Highly organized societies can also be found among the fearsome army ants in South America, Africa, and Asia, especially the genus *Dorylus*, of which a single colony may contain 20 million workers. Another behavioral convergence between ants and humans is encountered in the slave-making ant (genus *Polyergus*, widespread across the northern hemisphere), which raids the nests of closely related species to capture pupae, which are then raised in its own nest to become its workers.[55] Apparently, even human vices are paralleled among invertebrates.

Noting that the use of tools has emerged independently a number of times in the animal kingdom, Conway Morris contends that technology is inherent in evolution, as is intelligence, since tools, after all, epitomize intelligence and purpose. The use of tools by various bird species is well-known, but none is more impressive than the New Caledonian crow (*Corvus moneduloides*), which has a more sophisticated tool culture than chimpanzees. This includes the ability to make hooks from wood or fern, in an imposition of form that resembles carving. Once capability and motivation are combined, says Conway Morris, tool use becomes inevitable. This phenomenon also occurs among some New World monkeys, for example the capuchin monkey and the golden lion tamarin, and even with a species of parasitic wasp. That tool use is not a recent invention is suggested by the fossils of the bipedal ape *Oreopithecus*, which had a hand with a precision grip and lived in Tuscany between 7 and 3 million years ago.[56] We would add that the phenomenon of evolutionary convergence in this regard, including among birds and invertebrates, undermines the Darwinian belief that *Homo sapiens* is descended from ape-like "ancestors" such as *Homo habilis*, even if the latter's ability to use tools suggests some degree of intelligence and purpose.

55. Berg, *Nomogenesis*, 311; Conway Morris, *Life*, 198–201; Cooke et al., *Animals*, 578.
56. https://en.wikipedia.org/wiki/New_Caledonian_crow; Conway Morris, *Life*, 261–64, 269–70.

Convergence and Evolution

Evolutionary convergence is related to the phenomenon of equipotentiality, according to Conway Morris. The herpetologist Nick Arnold defines equipotentiality as follows:

> Different lineages of organisms often evolve a number of similar traits independently, but the order in which these are assembled may often be different, even in ecological analogues, especially if the taxa concerned are not closely related.... This [is the] phenomenon of equipotentiality, where more or less the same overall condition is reached by different routes.[57]

Consequently, as Conway Morris writes, evolution may not be fully inevitable, but it is highly predictable. It is also recognized, he adds, that evolution shows reversals instead of a linear, irreversible progression, but nonetheless the various biological properties (such as adaptability, cell structure, fertility, photosensitivity, and pigmentation) still arise.[58]

Although accepting that Neo-Darwinian random mutation provides the raw material upon which natural selection acts, Denis Alexander concurs on the importance of constraints and convergence in the evolutionary process. The British molecular biologist writes:

> So the rolling of the genetic dice in evolution is a wonderful way of generating both novelty and diversity, but at the same time it appears to be restrained by necessity to a limited number of living entities. If we live in a universe with this kind of physics and chemistry, and on a planet with these particular properties, then what we see is what we are likely to get ... only a relatively tiny number of genomes will generate organisms that can flourish in the different ecological niches on planet earth, and the evolutionary process keeps "finding" these again and again.... Evolutionary history on this planet displays overall increased complexity, genomic constraint and convergence.[59]

---

57. Quoted in Conway Morris, *Life*, 145.
58. Conway Morris, *Life*, 145, 223.
59. Alexander, *Creation*, 327, 330.

Convergent Evolution

Conway Morris concludes that convergence confirms the reality of organic evolution. While emphasizing the priority of adaptation (which we suggest occurs in a purposeful manner on the cellular level, as discussed earlier) in the evolutionary process, the English scientist approaches Hellenic metaphysics with the following statement: "The organic world is a plenitude and a marvel, but it still has a rational structure. Simplification will arise in its own adaptive circumstances, but so, too, will complexity."[60] In referring to the plenitude of the organic world, Conway Morris evokes Plato and the chain of being; in his mention of marvel, he calls to mind Aristotle, who, in his *Metaphysics*, declared that all philosophy begins in wonder; and in affirming the rational structure of the world, he recalls the numerical cosmology of Pythagoras as well as the Neoplatonic doctrine that the cosmos is grounded in the divine Logos, who sustains the world through the reason-principles (*logoi*) that indwell all things, including plants, animals, and humans, and provides them with intelligibility.

---

60. Conway Morris, *Life*, 301–03.

Conclusion

BEFORE TIME BEGINS, there is eternity. Before the many, there is the One, which in its essence is eternally beyond being and non-being. The One is the Principle of all manifestation, both directly (the intelligible world) and indirectly (the sensible world). According to the Hellenic metaphysical tradition that we have briefly surveyed, especially in its Neoplatonic culmination, the One produces Intellect (*nous*) out of the abundance of its goodness, in order to make manifestation possible. In its turn, Intellect produces Soul (*psychē*) as the means for interacting with matter, thereby giving rise to the cosmos. The physical cosmos arises out of pre-cosmic chaos through the activity of the World-soul, which is refracted into myriads of individual souls, namely, those of the celestial beings (gods/angels), humans, animals, and plants.

The active dimension of Intellect is the Logos, through which the One establishes the universe. As Heraclitus teaches, the Logos as cosmic lawgiver regulates the conflict between opposites that characterizes the world. The totality of the *logoi*, or reason-principles, through which the Intellect indwells the realm of manifestation, and on account of which the physical world possesses both being and intelligibility, are contained within the Logos. The Logos, called the Demiurge by Plato and the Prime Mover by Aristotle, creates by fashioning the cosmos according to numbers and geometrical figures, and accordingly, as the Pythagoreans taught, the cosmos arises in a sequence of numbers, lines, surfaces, and solids—that is to say, in a directed progression from non-dimensional numbers through one-dimensional lines and two-dimensional planes into three-dimensional bodies.

All solid bodies, including those of living beings, come to be through the interaction among a set of causes, of which four kinds are enumerated by Aristotle: (i) the formal cause provides the pattern or paradigm (*paradeigma*) of a thing; (ii) the material cause is

Conclusion

the particular kind(s) of matter (*hylē*) of which a thing is composed; (iii) the efficient cause is the agent of change or movement; and (iv) the final cause is the end or purpose (*telos*) for the sake of which a thing comes to be. Aristotle teaches that living beings have various levels of soul that determine their natures. Plants possess a nutritive or vegetative soul, which enables nutrition, growth and reproduction; animals possess the foregoing as well as a sensitive soul, which enables sense-perception, memory and imagination; and humans possess the foregoing as well as a rational soul, which enables reasoning (*dianoia*) and intellectual thought (*noēsis*). The rational soul is moreover the seat of immortality, inasmuch as it chooses to orient itself upwards to the realm of the eternal, instead of immersing itself in the material realm.

Since organic form (*morphē*) is numerical and geometrical in its foundation, it follows that any autonomic (i.e., regulated from within) transformation has to comply with natural laws that are mathematical in nature. We have presented the uniquely insightful work of D'Arcy Wentworth Thompson on the transformation of an impressive variety of organic forms to substantiate this argument, thereby affirming the Pythagorean and Platonic dictum that God is always geometrizing. An organic form can therefore only be transformed into another organic form in accordance with these laws, so that the process is called nomogenesis, or evolution according to natural law, as Lev Berg has persuasively argued.

We contend that nomogenesis pertains eminently to macro-evolution, which depends upon the generation of morphological novelty. This latter requires the production of new genetic material, thereby giving rise to forms that are substantially different from their predecessors. In this way complex structures, such as the eye or the feather, arise by means of a macro-mutational jump, instead of the gradual accumulation of minute variations postulated by modern evolutionary theory. This macro-evolutionary process also accounts for the introduction of new phyla, classes, orders, families, and probably also genera, of plants and animals. Macro-evolution thus fulfils the requirement of formal and final causality, against the rejection thereof in favor of mechanism and materialism by exponents of the modern scientific paradigm.

A clear distinction must be made between macro-evolution and micro-evolution, since the former is not merely the temporal extension of the latter, driven by the same mechanisms of random mutation and natural selection, as is claimed by Neo-Darwinism. This distinction depends on different mechanisms at the cellular level: whereas micro-evolution is driven by incremental cellular adaptations such as transposition, horizontal gene transfer, and epigenetics, macro-evolution is driven by the mutational "jumps" provided by symbiogenesis and genome duplication.

Micro-evolution is the process whereby variations arise within a species, and probably also species within a genus. These variations involve minor morphological changes, thus conforming to Aristotle when, defining the term "genus," he discusses the essential differences between one species and another as primarily due to differences of excess and defect. That is to say, species of the same genus differ mainly in relative magnitude and not in essential characters.[1] We contend that it is precisely this phenomenon that enables the participation of several species in the same genus. Pertinent examples among vertebrates are the genera *Panthera* (big cats), *Canis* (wolves/jackals), *Ursus* (bears), *Equus* (horse/zebra), *Cervus* (deer), *Cercopithecus* (monkeys), *Aquila* (eagles), *Phalacrocorax* (cormorants), *Crocodylus* (crocodiles), *Varanus* (monitors), *Rana* (frogs) and *Acipenser* (sturgeon). In all these cases the species comprising the particular genus differ mainly in relative magnitude, as well as adaptive characters such as coloration, and could therefore have become differentiated from each other by means of micro-evolution.

Scientists have established that for micro-evolution to occur, no new genetic material is required. Instead, the separation and recombination of existing genes are sufficient to produce such minor changes as bodily "excess" or "defect." In other words, micro-evolution entails genetic reshuffling and not genetic novelty, as in the case of macro-evolution. The Neo-Darwinian mechanisms of random mutation and natural selection are therefore sufficient for the production of new varieties and ultimately new species, at least on

---

1. Thompson, *Growth*, 273.

Conclusion

the empirical level; holistically speaking, all evolution involves the unfolding of that which has been in-folded. A further difference between macro- and micro-evolution is that the latter takes place gradually over numerous successive generations, whereas macro-evolution occurs relatively suddenly. Although the fossil record indicates that it is possible for a species and even a genus to evolve out of an existing one through gradual transformation, there is no evidence that a new family has ever arisen in this way.[2] Micro-evolution can account for the appearance of variations within a species and probably also species within a genus, but not for any of the taxa from the level of families upwards.

We have shown by means of numerous examples from the living kingdoms that the macro-evolutionary process is characterized by law, direction, and convergence. As a matter of fact, convergence is interlinked with orthogenesis or directed evolution, since the former results from development in a determined direction.[3] Furthermore, modern biochemistry demonstrates that biological macro-molecules cannot arise by trial and error, as happens when natural selection acts on the morphological level. This is confirmed by the paleontological evidence, which indicates that new groups corresponding to new genetic material arise suddenly and not gradually.[4] As a result of these biochemical and paleontological discoveries, we may confidently state that Darwinism is no longer entitled to its monopoly in evolutionary theory.

Against the Neo-Darwinian insistence that the evolutionary process is driven by random mutations, we contend that most mutations are purposeful adaptations occurring on the cellular level. We therefore agree with Perry Marshall that the modern theory of evolution violates fundamental engineering principles, for example information entropy. This concept is derived from information theory, which defines information entropy as the degree of uncertainty of a transmitted message caused by noise, where noise is the addition of randomness to a signal due to extraneous factors, such as

2. Dewar, *Illusion*, 262–63.
3. Berg, *Nomogenesis*, 157.
4. Swift, *Evolution*, 382–83.

occurs in radio interference from the sun. Noise thus has the same effect on electronic receivers that radiation has on DNA, namely the distortion of signals leading to a loss of information. In contrast, an authentic, post-Darwinian theory of evolution recognizes that cells actually reverse information entropy by communicating with each other, editing their genomes, exchanging DNA with other cells, engaging in symbiotic relationships, and forming hybrids through genome duplication. Through these adaptive mechanisms, cellular behavior increases information and order in the universe. Moreover, there is increasing evidence from molecular biology that cellular adaptation is not random but goal-directed. Ultimately, the random mutation hypothesis of Neo-Darwinism stands in opposition to the scientific method itself, because randomness cannot be mathematically verified.[5]

In order to accommodate an alternative evolutionary theory, Perry Marshall suggests a new master paradigm for the life sciences to supplement the current paradigm for the physical sciences. According to the latter, the phenomena of matter, energy, space, and time are organized according to discoverable laws of physics, while these laws are viewed as immutable and universal. According to the new biological paradigm that Marshall proposes, living things always obey the universal laws of physics and chemistry. These laws, however, cannot exhaustively explain the behavior of living things, because of the subjectivity and intentionality of each living being. In addition, it is affirmed that all life is based on codes, but in contrast to the physical and chemical laws, the laws of code are not unalterable and universal, but chosen and local. Moreover, it is recognized that each layer of code bestows a higher level of intent, so that the code-guided behavior of living things is ultimately teleological.[6] In this way, through scientific argumentation, we arrive more or less at our point of departure, namely the Hellenic notions of formal and final causality.

The case for an alternative theory of evolution that recognizes formal and final causality has in recent decades been powerfully

---

5. Marshall, *Evolution*, 249, 259, 284–85, 292.
6. Ibid., 214, 216.

Conclusion

argued by scientists such as Michael Denton and David Swift. As Denton writes, "The possibility that life on earth approximates to the plenitude of all possible biological forms is perfectly in keeping with the teleological thesis that the cosmos is uniquely prefabricated for life as it exists on earth"; and: "Four centuries after the scientific revolution apparently destroyed irretrievably man's special place in the universe, banished Aristotle, and rendered teleological speculation obsolete, the relentless stream of discovery has turned dramatically in favor of teleology and design."[7] Nonetheless, atheistic scientists and others who reject final causality are still insisting that the evolutionary process is entirely accidental and purposeless. Thus Richard Dawkins writes, "Evolution produces such a strong illusion of design it has fooled almost every human who ever lived." Dawkins' statement reminds us of Alfred North Whitehead's witticism, "Those who devote themselves to the purpose of proving that there is no purpose constitute an interesting subject for study."[8]

The relation between formal causality and design has been usefully summarized by Perry Marshall, who compares the living and non-living realms. Both a watch and a tree are built according to a set of instructions, namely the information in its plan and the instructions in its DNA, respectively. In both cases, an idea (i.e., the formal cause) precedes its embodiment, which is precisely what design entails. Consequently, "watches and trees are ultimately a product of design."[9] Facts and arguments such as the foregoing led Simon Conway Morris to admit,

> No wonder the arguments for design and intelligent planning have such a perennial appeal. Whether it be by navigation across the hyperdimensional vastness of protein space, the journey to a genetic code of almost eerie efficiency, or the more familiar examples of superb adaptation, life has an extraordinary propensity for its metaphorical hand to fit the glove.... Its central paradox revolves around the fact that despite its fecundity and baroque richness life is also strongly constrained. The net result is a genu-

---

7. Denton, *Destiny*, 299, 389.
8. Both quoted in Marshall, *Evolution*, 245.
9. Marshall, *Evolution*, 217.

ine creation, almost unimaginably rich and beautiful, but one also with an underlying structure in which, given enough time, the inevitable must happen.[10]

Metaphysically speaking, the appearance of a new organic form through mutation, as in macro-evolution, occurs in accordance with a morphological plan or structure which is the formal cause, while its survival through reproduction is the final cause or purpose of phylogenetic development. That is to say, the indwelling presence of the *logoi* ensures both formal and final causality in the organic realm. In this manner the physical reflects the metaphysical, from which it cannot be divorced without a radical loss of both ontological depth and epistemological scope.

---

10. Conway Morris, *Life*, 19–20.

Bibliography

Alexander, Denis. *Creation or Evolution. Do We Have to Choose?* Oxford, UK & Grand Rapids, MI: Monarch Books, 2008.

Aquinas, Thomas. *Selected Philosophical Writings.* Translated by Timothy McDermott. Oxford & New York: Oxford University Press, 1993.

Aristotle. *Metaphysics.* Translated by W.D. Ross. In *The Basic Works of Aristotle*, edited by Richard McKeon. New York: The Modern Library, 2001.

_____. *On Generation and Corruption.* Translated by Harold J. Joachim. In *The Basic Works of Aristotle*, edited by Richard McKeon. New York: The Modern Library, 2001.

_____. *On the Generation of Animals.* Translated by Arthur Platt. In *The Basic Works of Aristotle*, edited by Richard McKeon. New York: The Modern Library, 2001.

_____. *On the Heavens.* Translated by J.L. Stocks. In *The Basic Works of Aristotle*, edited by Richard McKeon. New York: The Modern Library, 2001.

_____. *On the Parts of Animals.* Translated by William Ogle. In *The Basic Works of Aristotle*, edited by Richard McKeon. New York: The Modern Library, 2001.

_____. *On the Soul.* Translated by J.A. Smith. In *The Basic Works of Aristotle*, edited by Richard McKeon. New York: The Modern Library, 2001.

_____. *Physics.* Translated by R.P. Hardie and R.K. Gaye. In *The Basic Works of Aristotle*, edited by Richard McKeon. New York: The Modern Library, 2001.

_____. *The History of Animals.* Translated by D'Arcy Wentworth Thompson. In *The Basic Works of Aristotle*, edited by Richard McKeon. New York: The Modern Library, 2001.

Bailey, Jim. *Sailing to Paradise. The Discovery of the Americas by 7000 BC.* New York: Simon & Schuster, 1994.

Bakar, Osman. "The Nature and Extent of Criticism of Evolutionary Theory," in *Science and the Myth of Progress*, 2003. Edited by Mehrdad Zarandi. www.worldwisdom.com/public/library/default.aspx.

Berg, Leo. *Nomogenesis, or Evolution Determined by Law*. Translated by J.N. Rostovtsov. Cambridge, MA and London, England: MIT Press, 1969.

Bergman, Jerry. *The Rise and Fall of the Orthogenesis Non-Darwinian Theory of Evolution*, 2009. https://www.creationresearch.org/crsq/articles/47/47_2/CRSQ%20Fall%202010%20Bergman.pdf.

Blackburn, Simon. *The Oxford Dictionary of Philosophy*. Oxford: Oxford University Press, 2008.

Burckhardt, Titus. "Cosmology and Modern Science," in *The Sword of Gnosis. Metaphysics, Cosmology, Tradition, Symbolism*. Edited by Jacob Needleman. Baltimore, MD: Penguin Books, 1974.

Bynum, William F. "The Great Chain of Being after forty years: an appraisal," in *History of Science*, Vol 13, 1975, 1–28. http://adsabs.harvard.edu/abs/1975HisSc..13....1B.

Carabine, Deidre. *John Scottus Eriugena*. New York and Oxford: Oxford University Press, 2000.

Conway Morris, Simon. *Life's Solution. Inevitable Humans in a Lonely Universe*. Cambridge: Cambridge University Press, 2003.

Cooke, Fred; Dingle, Hugh; Hutchinson, Stephen; McKay, George; Schodde, Richard; Tait, Noel; Vogt, Richard (editors). *The Encyclopedia of Animals. A Complete Visual Guide*. Sydney: Weldon Owen, 2008.

Coomaraswamy, Ananda K. *What is Civilization? And Other Essays*. Ipswich: Golgonooza Press, 1989.

Cooper, John M., editor. *Plato. Collected Works*. Indianapolis, IN: Hackett Publishing, 1997.

Cornford, Francis. *Plato's Cosmology. The Timaeus of Plato*. Indianapolis, IN: Hackett Publishing, 1997.

Critchlow, Keith. Foreword to *Quadrivium. The four classical liberal arts of number, geometry, music & cosmology*. Edited by John Martineau. Glastonbury: Wooden Books, 2010.

Curd, Patricia. *Presocratic Philosophy*. Stanford Encyclopedia of Philosophy, 2011. http://plato.stanford.edu/entries/presocratics.

Cutsinger, James S. "On Earth as it is in Heaven," 2007. http://www.cutsinger.net/pdf/ earth_as_it_is_in_heaven.pdf.

Darwin, Charles. *The Origin of Species*. Ware, Hertfordshire: Wordsworth, 1998.

Davison, John A. *An Evolutionary Manifesto. A New Hypothesis for Organic Change*, 2000. http://www.uvm.edu/~jdavison/davison-manifesto.html#order.

_____. *The case for instant evolution*, 2005. http://www.iscid.org/papers. Davison_InstantEvolution_050204.pdf.

_____. *A Prescribed Evolutionary Hypothesis*, 2006. http://www.evcforum.net/DataDropsite/APrescribedEvolutionaryHypothesis.html.

Dawkins, Richard. *The Ancestor's Tale. A Pilgrimage to the Dawn of Life*. London: Weidenfeld & Nicolson, 2004.

Denton, Michael. *Nature's Destiny. How the Laws of Biology Reveal Purpose in the Universe*. New York: The Free Press, 1998.

Denton, Michael J., Marshall, Craig J. and Legge, Michael. "The Protein Folds as Platonic Forms. New Support for the pre-Darwinian Conception of Evolution by Natural Law," in *Journal of Theoretical Biology*, 219 (2002): 325–342.

Dewar, Douglas. *The Transformist Illusion*. Hillsdale, NY: Sophia Perennis, 2005.

Dillon, John and Gerson, Lloyd (editors). *Neoplatonic Philosophy. Introductory Readings*. Indianapolis, IN: Hackett Publishing, 2004.

Dobzhansky, Theodosius. *The Biology of Ultimate Concern*. London: Fontana, 1971.

_____. "Nothing in biology makes sense except in the light of evolution," in *The American Biology Teacher*, 35 (1973): 125–129. http://www.pbs.org/wgbh/evolution/library/10/2/text_pop/l_102_01.html.

Dreyer, P.S. *Die Wysbegeerte van die Grieke* [The Philosophy of the Greeks]. Kaapstad & Pretoria: HAUM, 1975.

Ferguson, Kitty. *Pythagoras. His Lives and the Legacy of a Rational Universe*. London: Icon Books, 2011.

Flannery, Michael A. *Alfred Russell Wallace's Theory of Intelligent Evolution*. Riesel, TX: Erasmus Press, 2011.

Fuller, Steven. *Dissent over Descent. Intelligent Design's Challenge to Darwinism*. Cambridge: Icon Books, 2008.

Gerson, Lloyd P. *Aristotle and Other Platonists*. Ithaca, NY: Cornell University Press, 2005.

Gilson, Etienne. *From Aristotle to Darwin and Back Again. A Journey in Final Causality, Species, and Evolution*. Translated by John Lyon. San Francisco: Ignatius Press, 2009.

Goosen, Danie. *Die Nihilisme. Notas oor ons tyd* [Nihilism. Notes on our time]. Pretoria: Praag, 2007.

Gould, Stephen Jay. "Darwinian Fundamentalism," in *The New York Review of Books* (Volume 44, Number 10, 1997).

_____. *The Richness of Life: The Essential Stephen Jay Gould*. Edited by Paul McGarr and Steven Rose. London: Vintage Books, 2007.

Griffin, David Ray. "Interpreting Science from the Standpoint of Whiteheadian Process Philosophy," in *The Oxford Handbook of Religion and Science*. Edited by Philip Clayton and Zachary Simpson. Oxford: Oxford University Press, 2008.

Heidegger, Martin. *Introduction to Metaphysics*. Translated by Gregory Fried and Richard Polt. New Haven: Yale University Press, 2000.

Hewlett, Martinez. "Molecular Biology and Religion," in *The Oxford Handbook of Religion and Science*. Edited by Philip Clayton and Zachary Simpson. Oxford: Oxford University Press, 2008.

Iamblichus. *The Theology of Arithmetic. On the Mystical, Mathematical and Cosmological Symbolism of the First Ten Numbers*. Translated by Robin Waterfield. Grand Rapids, MI: Phanes Press, 1988.

Liddell and Scott Greek-English Lexicon (Abridged Edition). Oxford: Oxford University Press, 2004.

Lings, Martin. "Signs of the Times," in *The Sword of Gnosis. Metaphysics, Cosmology, Tradition, Symbolism*. Edited by Jacob Needleman. Baltimore, MD: Penguin Books, 1974.

Lossky, Vladimir. *Orthodox Theology. An Introduction*. Translated by Ian and Ihita Kesarcodi-Watson. Crestwood, NY: St Vladimir's Seminary Press, 1978.

Lovejoy, Arthur O. *The Great Chain of Being. A Study of the History of an Idea*. New York: Harper & Brothers, 1960.

Lundy, Miranda. "Sacred Number" and "Sacred Geometry," in *Quad-*

rivium. The four classical liberal arts of number, geometry, music & cosmology. Edited by John Martineau. Glastonbury: Wooden Books, 2010.

MacBride, E.W. "Berg's Nomogenesis. A criticism of natural selection," in *The Eugenics Review*, April 1927; 19(1):32–37. http://www.ncbi.nlm.nih.gov/pmc/articles/PMC2984689.

MacNeill, Allen. *The Platonic Roots of Intelligent Design Theory.* http://www.evolutionlist.blogspot.com/2006/02/platonic-roots-of-intelligent-design.html.

Marshall, Perry. *Evolution 2.0: Breaking the Deadlock between Darwin and Design.* Dallas, TX: BenBella Books, 2015.

Martineau, John. "A Little Book of Coincidence," in *Quadrivium. The four classical liberal arts of number, geometry, music & cosmology.* Edited by John Martineau. Glastonbury: Wooden Books, 2010.

McKeon, Richard, editor. *The Basic Works of Aristotle.* New York: The Modern Library, 2001.

McKirahan, Richard. *Philosophy before Socrates. An Introduction with Texts and Commentary.* Indianapolis, IN: Hackett Publishing, 1994.

Moore, Edward. *Neoplatonism*, in Internet Encyclopaedia of Philosophy, 2005. http://iep.utm.edu/neoplato.

_____. *Plotinus*, in Internet Encyclopedia of Philosophy, 2001. http://www.iep.utm.edu/plotinus.

Nasr, Seyyed Hossein. *Islamic Science. An Illustrated Study.* World of Islam Festival Publishing Company Ltd, 1976.

_____. "Progress and Evolution. A Reappraisal from the Traditional Perspective," in *Parabola* (Volume VI, Number 2, 1981).

Nietzsche, Friedrich Wilhelm. *Thus Spoke Zarathustra.* Translated by R.J. Hollingdale. Harmondsworth: Penguin Books, 1969.

_____. *Twilight of the Idols.* Translated by Antony Ludovici. London: Wordsworth, 2007.

Northbourne, Lord. *Looking Back on Progress.* Ghent, NY: Sophia Perennis et Universalis, 1995.

Oosthuizen, J.S. *Van Plotinus tot Teilhard de Chardin. 'n Studie oor die metamorfose van die Westerse werklikheidsbeeld* [From Plotinus to Teilhard de Chardin. A study on the metamorphosis of the Western world-view]. Amsterdam: Rodopi N.V., 1974.

O'Rourke, Fran. "Aristotle and the Metaphysics of Evolution," in *The Review of Metaphysics* 58 (September 2004): 3–59.

Plato. *The Republic*. Translated by Desmond Lee. London: Penguin Books, 1987.

———. *Theaetetus*. Translated by M. J. Levett, revised by Myles Burnyeat. In *Collected Works*, edited by John M. Cooper. Indianapolis, IN: Hackett Publishing, 1997.

———. *Timaeus*. Translated by Donald J. Zeyl. In *Collected Works*, edited by John M. Cooper. Indianapolis, IN: Hackett Publishing, 1997.

Popov, Igor. "The problem of constraints on variation, from Darwin to the present," in *Ludus Vitalis*, Vol. XVII, no. 32 (2009): 201–220. https://www.researchgate.net/...Popov2/...CONSTRAINTS_ON_VARIATION.../551b...

Ross, David. *Aristotle*. London and New York: Routledge, 1995.

Schuon, Frithjof. *Esoterism as Principle and as Way*. Translated by Willian Stoddart. Pates Manor, Middlesex: Perennial Books, 1981.

———. *From the Divine to the Human. Survey of Metaphysics and Epistemology*. Translated by Gustavo Polit and Deborah Lambert. Bloomington, IN: World Wisdom Books, 1982.

———. *Sophia Perennis and the theory of evolution and progress*, 2001. http://www.frithof-schuon.com/evolution-engl.htm.

Sheldon-Williams, I. P. "The Greek Christian Platonist Tradition from the Cappadocians to Maximus and Eriugena," in *The Cambridge History of Later Greek and Early Medieval Philosophy*, edited by A. H. Armstrong. London: Cambridge University Press, 1967.

Smith, Wolfgang. *Cosmos and Transcendence*. San Rafael, CA: Sophia Perennis, 2008.

Spengler, Oswald. *The Decline of the West* (Abridged Edition). Translated by Charles Atkinson. Edited by Helmut Werner. Oxford: Oxford University Press, 1991.

Sutton, Daud. "Platonic and Archimedean Solids," in *Quadrivium. The four classical liberal arts of number, geometry, music & cosmology*. Edited by John Martineau. Glastonbury: Wooden Books, 2010.

Swift, David. *Evolution under the Microscope. A Scientific Critique of the Theory of Evolution*. Stirling: Leighton, 2002.

Bibliography

Taylor, Thomas. *Introduction to the Philosophy & Writings of Plato*. Seaside, OR: Watchmaker Publishing, 2010.

Theodossiou, Efstratios, Manimanis, Vassilios N. and Dimitrijevic, Milan S. "The cosmological theories of the pre-Socratic Greek philosophers and their philosophical views for the environment," in *Facta Universitatis*, Vol. 10, No. 1, 2011, 89–99.

Thompson, D'Arcy Wentworth. *On Growth and Form* (Abridged edition). Cambridge and New York: Cambridge University Press, 1992.

Thompson, Richard L. *Mechanistic and Nonmechanistic Science. An Investigation into the Nature of Consciousness and Form*. Los Angeles: The Bhaktivedanta Book Trust, 1981.

_____. *Origins. Higher Dimensions in Science*. Mumbai: The Bhaktivedanta Book Trust, 1984.

Tkacz, Michael W. "Thomistic Reflections on Teleology and Contemporary Biological Research," in *New Blackfriars* 94 (2013): 654–75.

Uzdavinys, Algis. *The Golden Chain. An Anthology of Pythagorean and Platonist Philosophy*. Bloomington, IN: World Wisdom Books, 2004.

_____. *Orpheus and the Roots of Platonism*. London: The Matheson Trust, 2011.

Van Till, Howard J. "Basil, Augustine, and the Doctrine of Creation's Functional Integrity." *Science & Christian Belief* 8, No. 1 (1996): 21–38. http://www.asa3.org/ASA/topics/Evolution/S&CB4-96VanTill.html.

_____. "Basil and Augustine Revisited: The Survival of Functional Integrity," *Origins & Design*, 19:1 (1998). http://www.arn.org/docs/odesign/od191/basilaug191.htm.

Van Vrekhem, Georges. *Evolution, Religion and the Unknown God*. CreateSpace Independent Publishing Platform, 2012.

Vlastos, Gregory. *Plato's Universe*. Oxford: Clarendon Press, 1975.

Ward, Keith. *The Big Questions in Science and Religion*. West Conshohocken, PA: Templeton Foundation Press, 2008.

Yannaras, Christos. *Postmodern Metaphysics*. Translated by Norman Russell. Brookline, MA: Holy Cross Orthodox Press, 2004.

Yockey, Francis Parker. *Imperium. The Philosophy of History and Politics*. Torrance, CA: The Noontide Press, 1962.

From Logos to Bios

Zeyl, Donald. *Plato's Timaeus*. Stanford Encyclopedia of Philosophy, 2009. http://plato.stanford.edu/entries/plato-timaeus.

Index

Adaptation 115, 117–18, 137, 149–52, 161–62, 167–68, 197, 199, 208, 210–13, 216, 225, 252, 257, 262
Adaptive radiation 170–71, 202
Alexander, Denis 90, 133–34, 164, 175, 185–89, 219, 235, 237, 249, 256
Anaxagoras 20–22, 40–41
Anaximander 15–16, 39–40
Aquinas, Thomas 8, 45–47, 64
Aristotle 10–11, 27–32, 39–41, 44–60, 62, 67, 71–72, 129, 134, 142, 150, 154–55, 164, 176, 194–95, 236, 257–260
Arnold, Nick 256
Artificial breeding/selection 121, 127–28, 145, 158–59, 209
Augustine of Hippo 7
Avicenna 63

Bakar, Osman 6, 77, 99, 105, 132–33, 140, 142
Basil of Caesarea 7, 70
Bateson, William 108, 172
Beauty 27, 35, 57–58, 80, 86, 120, 135, 137, 226
Behe, Michael 233–34
Belon, Pierre 236
Bentham, Jeremy 136
Berg, Lev 88, 100, 127, 134–35, 149–54, 158–69, 172–74, 178–84, 202–08, 211, 214, 235, 239–41, 245–48, 251–52, 255, 259
Bergman, Jerry 198, 202, 211, 215, 248
Berlinski, David 119
Bjerring, Hans 94
Bohm, David 153
Bounoure, Louis 125
Brethren of Purity 63
Broom, Robert 110–11
Buffon, Comte de 177
Burbank, Luther 209
Burckhardt, Titus 6, 121, 132, 134, 177, 188, 229

Calvin, Calvinism 134, 138
Cambrian Explosion 113–14, 116, 184–85, 187
Camper, Peter 80
Chambers, Robert 9, 99, 146–47
Chance, contingency 30–31, 59, 115–17, 128–32, 148, 151–56, 168, 182–83, 209, 245
Chain of Being 1, 3, 6, 62–70, 257
Chaos theory 131
Chomsky, Noam 253
Complexity 45, 56, 66, 94, 122–23, 132–33, 154–57, 200, 203, 207, 231–34, 256–57
Code 56, 130–31, 143, 166, 184, 217–22, 262
Constraints 80, 84, 104, 117, 193, 198, 207–12, 226, 228, 231–34, 245, 256
Conway Morris, Simon 116–17, 120, 192, 214–15, 237–38, 241–250
Cooke, Fred 83, 89–92, 96, 151,

273

164, 167–70, 185–202, 211, 214,
235, 241–42, 245–51, 255
Coomaraswamy, Ananda i, 2, 6–
12, 23, 25
Coyne, Jerry 119, 156
Creation, Divine 1, 6, 105, 111, 131
Creationism 27, 99, 105, 146
Crick, Francis 156, 218
Cutsinger, James 5, 7, 12–13, 40,
70, 144
Cuvier, Georges 105, 226, 233

Danilevsky, Nikolai 120–21
Darwin, Charles 1, 87–88, 99–109,
113–37, 141, 145, 148–54, 159, 161,
163, 171–74, 178, 180, 194, 200,
205–08, 222, 226, 232–35, 247
Darwin, Erasmus 236
Darwinism, Darwinian 1–3, 13,
47, 54, 56, 59–60, 68, 89, 92,
97–101, 105, 108–12, 115–40,
143–44, 148–49, 154–60, 166,
169, 171, 175–77, 182–83, 185–92,
196–98, 201, 204–10, 213–20,
223–25, 227, 230–36, 245–46,
252, 255–56, 260–62
Davies, Paul 69
Davison, John 94, 97, 109, 111,
121–22, 153, 157, 161, 163–66, 181,
203, 209, 210, 237, 241, 249
Dawkins, Richard 115–19, 131, 148,
171, 178, 218–220, 234, 263
De Beer, Gavin 246
Delbrück, Max 216
Demiurge 3, 24–27, 32, 42–43, 67,
258
Denton, Michael 44–45, 72, 102,
113, 115, 146–47, 148, 192–95,
197, 200, 206, 213, 215–19, 226–
34, 263

Dennett, Daniel 102, 117, 154
Descartes, René 1, 78–79, 139
De Vries, Hugo 108, 167, 179–80
Dewar, Douglas 84, 87–88, 90–97,
124, 127–28, 133, 173, 185–90,
230, 240, 242, 247, 261
Dilley, Stephen 148
Diogenes 13, 22
DNA 41, 48–50, 59, 74, 111, 129–31,
154–56, 176, 213, 215–21, 223–24,
228, 231, 237, 247, 262
Dobzhansky, Theodosius 109–12,
129, 140–41, 149, 159–60, 169–
70, 199, 206, 239
Dodds, E. R. 11
Dürer, Albrecht 80

Eden, Murray 128–29
Eimer, Theodor 198
Einstein, Albert 143
Eldredge, Niles 94, 112–13, 224
Empedocles 21, 36–37, 41–42, 59,
150
Entropy 133–34, 154, 161, 182, 184,
261–62
Epigenetics 220, 222–25
Equipotentiality 256
Eriugena, John Scottus 63, 68
Essentialism 10–11, 178
Eusebius 16
Extinction 116–17, 148, 161, 163–
65, 168, 184, 208, 210–11

Fibonacci sequence 75
Filipchenko, Yuri 212
Fisher, Ronald 108
Flannery, Michael 45, 101, 105,
119, 127, 129, 137, 154, 201
Fossil record, fossils 81, 83, 87–97,
113–14, 118, 123–26, 179–81, 185–

Index

91, 200, 202, 207, 210, 212, 214, 219, 237, 248, 255, 261
Fuller, Steven 57, 103, 107, 115, 129, 137

Gerson, Lloyd 26–27, 31–32, 34, 37, 46, 48, 50–51, 61, 68, 177
Gilson, Etienne 5, 10, 15, 55, 57, 59, 99, 101, 105, 119, 128, 137
Golden Ratio 75–76
Goldschmidt, Richard 109, 122, 165, 181
Gould, Stephen Jay 76, 96, 112–17, 207–08, 212–13, 224
Gradualism 98–99, 115, 177–78, 186
Grassé, Pierre 110, 121, 153, 161, 203
Gray, Patricia 254
Gribbin, John 164
Guénon, René i
Guttman, Burton 124

Haacke, Wilhelm 198
Haeckel, Ernst 119, 125, 221
Haldane, John 60, 108
Hawking, Stephen 100
Hegel, Georg 138, 141
Heidegger, Martin 19
Heraclitus 18–20, 23, 26, 258
Heredity 10, 49, 102, 107–08, 158, 180, 198, 200, 216
Heribert-Nilsson, Nils 88
Hewlett, Martinez 111, 156, 214, 218
Holistic 69, 194, 210, 213, 232, 234, 261
Holmes, Oliver 80
Homer 62
Homology 53, 91, 205–06, 229, 235–36, 242, 245–46
Horizontal gene transfer 220–21, 225, 227, 234, 260

Hoyle, Fred 132, 156
Huxley, Julian 110, 140
Huxley, Thomas 82, 117
Hybridization (genome duplication) 118–19, 220, 224–25

Iamblichus 35, 71–74
Information theory 132, 154, 161
Intellect/Mind 3–4, 15, 20–25, 27, 33–34, 37, 44, 51–52, 55, 61, 63, 69, 73, 132, 137, 142, 144, 258

Jarvik, Erik 94
Jollos, Viktor 205
Junk hypothesis 218–19

Kant, Immanuel 141, 152, 177, 228
Kauffman, Stuart 232–33
Kenyon, Dean 196
Koestler, Arthur 206
Kovalenko, Elena 214
Kozo-Polyansky, Boris 223
Kullmann, Wolfgang 50

Lamarck, Jean-Baptiste 99, 103, 107, 222–23
Language 143, 174–74, 219, 221–22, 253
Lanza, Robert 143
Larson, Edward 101
Lee, Desmond 22–23
Leibniz, Gottfried 66, 180
Lemoine, Paul 10, 173
Lewontin, Richard 100, 162, 193, 212
Lings, Martin 122, 139–40
Linnaeus, Carl 53, 65–66
Living fossils 126, 202
Logos, *logoi* 18–20, 23, 31, 34, 56, 59–60, 63, 65–66, 257–58

Lovejoy, Arthur 33, 65–68
Lundy, Miranda 18, 28, 72–74
Luria-Delbrück experiment 216
Lyell, Charles 118, 136

MacBride, Ernest 159, 180
Macro-evolution 2–4, 56, 78, 94, 108, 110–11, 121–25, 127, 132, 145, 154, 157, 165, 179, 181, 184, 191, 212, 225, 259–61, 264
Malthus, Thomas 118, 134–35
Margulis, Lynn 223
Marino, Lori 253
Marshall, Craig 44, 192, 213, 215
Marshall, Perry 28, 69, 76, 88–89, 98, 100, 102, 108–10, 112, 114–15, 119–20, 129–36, 139, 141, 143, 148, 153–56, 161, 176, 184–85, 216–217, 219–28, 233–34, 261–63
Marx, Karl 139, 141
Materialism 1–2, 66, 119, 141, 143, 252, 259
Mayr, Ernst 122, 175, 178, 206, 209
McClintock, Barbara 220
McKirahan, Richard 13, 16–22, 40–41
Meillet, Antoine 174–75
Mendel, Gregor 106–08, 110, 127
Mereschowsky, Konstantin 223
Micro-evolution 3, 78, 95, 108, 110, 121–23, 125, 127, 132, 145, 157, 162, 165, 170, 175, 179–81, 191, 199–200, 212, 222, 225, 260–61
Mill, John Stuart 136–37, 53
Mind (see Intellect)
Molecular biology 1, 41, 49–50, 56, 123, 155–57, 191, 213, 218, 220, 225, 262
Molecular convergence 238
Morgan, Thomas 108–09

Morphology 59, 76–78, 93, 174–75, 193, 197, 200, 221, 230, 235, 239, 249, 251

Nagel, Thomas 112
Nasr, S. H. 1–2, 63–64, 138–40
Natural law 16, 40–41, 45, 86, 100, 146–49, 152–53, 157, 182, 192–93, 196, 209, 235, 259
Natural theology 154, 157
Necessity 3–4, 15–16, 25–27, 31, 56, 59–60, 67–68, 96, 123, 148, 256
Needham, Joseph 54
Neo-Darwinian, Neo-Darwinism 3, 56, 59–60, 68, 92, 98, 110–115, 128–131, 140, 143, 156–57, 160, 166, 175, 185–86, 190–91, 199, 206, 208, 214, 216, 218, 220, 225, 236, 252, 256, 260–62
Neoplatonist, Neoplatonism 3, 15, 18, 33–36, 38, 53, 60–62, 67, 73–74, 257–58
Nichols, Terence 59–60
Nietzsche, Friedrich 70, 135
Nilsson, Dan-Eric 88, 242
Nomogenesis 149, 153, 159, 182–83, 245, 259
Northbourne, Lord 58, 73, 135, 147, 153

Ohno, Susumu 218, 224
O'Rourke, Fran 44, 48–50, 52, 54–56, 60, 176
Orthogenesis 108, 198–202, 208, 210–11, 215, 261
Osborn, Henry 102, 201–02, 204
Owen, Richard 99, 147, 149

Paleontology, paleontological 1, 11, 90, 94–95, 102, 110, 112–13, 116–

Index

17, 120, 124–26, 134, 149, 161–62, 164, 172, 179–81, 184, 189–91, 201, 204, 207, 210, 212, 214, 237, 261
Parallelism 80, 149, 214, 235–36, 245–46
Parmenides 23, 37
Participation 35, 51, 62, 70, 260
Pelger, Susanne 242
Perez, Jean-Claude 76
Phylogenetic acceleration 151
Planck, Max 181
Plato, Platonism, Platonic 3–4, 9–12, 15, 17–19, 22–30, 32–34, 36–38, 40, 42–43, 45–46, 51–52, 60, 63, 66–68, 71, 73–74, 76–77, 134, 143, 147–48, 178, 192, 194–95, 197, 216, 236, 254, 257–59
Plenitude 66–67, 225–27, 230–31, 257, 263
Plotinus 27, 33–35, 37, 53, 60–61, 67–68
Plutarch 10, 71
Polanyi, Michael 218
Polyphyletic origins 90–91, 104, 171–73, 179, 183, 204, 239–40
Popov, Igor 104, 120, 121, 198–99, 200–01, 205–08, 210, 212, 214, 229, 236
Population genetics 108–09, 113, 214
Prime Mover 31–32, 55, 62, 258
Proclus 11, 36–37, 61–62
Progress 66, 100, 112, 114–15, 123, 138–40, 150, 199, 206
Protein folds 192–97
Psychical convergence 252–54
Punctuated equilibrium 112–13, 118, 132
Purpose 3, 5, 10, 15, 25, 31, 50–51, 55–56, 129, 131, 140, 151, 154–55,

158, 184, 255, 259, 263–64
Pythagoras, Pythagorean 3, 16–18, 40, 71–72, 76–77, 216, 257–59

Recapitulation 182
Receptacle of becoming 25–26, 30
Redundancy 213
Richard the Englishman 6

Schindewolf, Otto 94, 124, 162, 181, 210, 249
Schneider, Karl 171
Schopenhauer, Arthur 134
Schrödinger, Erwin 184
Schumacher, Ernst 68
Schuon, Frithjof i, 6–9, 30, 70, 73, 141–42, 145
Schützenberger, Marcel 128–29
Schwendener, Simon 223
Seilacher, Adolf 212–13
Sensory convergence 241–43
Shapiro, James 220, 224
Shute, Evan 122
Simpson, George 124, 190–91
Smith, Adam 101
Smith, John Maynard 117
Smith, Wolfgang 119, 125, 127, 181, 184, 191
Soul 3, 12–13, 15, 24, 28, 33–34, 37–38, 43–48, 50–55, 60–61, 67, 71, 73, 142, 258–59
Spencer, Herbert 99, 134
Spengler, Oswald 63, 66, 125–26, 136
Spurway, Helen 206
Stasis 112–13, 124, 131–32, 191
Steinman, Gary 196
Stoltzfus, Arlin 214
Swift, David 9–10, 28, 40, 42, 46, 55–56, 88–95, 101, 107–13, 118,

122–25, 130, 137, 145, 152, 154–56, 161–62, 166, 171, 176, 179, 185–87, 189, 191, 200, 207, 211, 217, 219, 229, 236–37, 242, 245–46, 252, 261–63
Symbiogenesis 88–89, 114, 119, 220, 223–25, 260, 262

Tkacz, Michael 58–59
Teleology, teleological 4, 10, 14–15, 21–22, 31, 49, 55–57, 59–60, 100, 112, 115, 136–37, 154–55, 184, 262–63
Theodossiou, Efstratios 16, 19, 39, 43, 61
Thompson, D'Arcy Wentworth 3, 14, 49, 59, 73, 76–86, 100, 102, 121, 149, 156, 191, 194, 209–10, 253, 259–60
Thompson, Richard 88, 105–06, 112, 123, 132–33 153, 155, 157–58, 189–90, 209
Thompson, William 137
Transposition 220–22, 225, 234, 260
Tree of life 125, 226, 231

Uniformitarianism 105, 115

Van Till, Howard 7–9, 11, 69, 131–32
Van Vrekhem, Georges 65–66, 68–69, 99, 101–102, 105, 107–10, 112–13, 115–16, 118–19, 121, 124, 162, 165, 193
Vavilov, Nikolay 204–05, 246
Von Bertalanffy, Ludwig 206

Waagen, Wilhelm 179
Wallace, Alfred 45, 99, 101, 127, 136–37, 200–01, 212
Weber, Max 138
Weismann, August 107, 120, 198, 212, 216, 222
Whitehead, Alfred North 2, 263
Whitman, Charles 201, 205
Wickramasinghe, Chandra 132, 156
Wiener, Norbert 143
Woese, Carl 227
Wright, Sewall 108

Yampolsky, Lev 214
Yannaras, Christos 142–43
Yockey, Francis Parker 63, 124, 126–27, 134–36

Zeising, Adolf 75–76

About the Author

Wynand de Beer is a South African who taught in Cape Town until moving to Ireland, where he met his Russian wife Lyudmila. His research articles on Hellenic philosophy and Patristic theology have been published in various peer-reviewed journals in the United States and South Africa. A member of the Russian Orthodox Church, he has also written articles for Orthodox publications and websites under his Orthodox name, Vladimir de Beer.

www.ingramcontent.com/pod-product-compliance
Lightning Source LLC
Chambersburg PA
CBHW030103170426
43198CB00009B/481